Tikhonov Vasil'eva Sveshnikov Differential Equations

A. N. Tikhonov   A. B. Vasil'eva   A. G. Sveshnikov

Differential Equations

Translated from the Russian
by A. B. Sossinskij

With 30 Figures

Springer-Verlag
Berlin Heidelberg New York Tokyo

Andrei N. Tikhonov
Keldysh Institute of Applied Mathematics of the
Academy of Sciences of the USSR

Adelaida B. Vasil'eva · Alexei G. Sveshnikov
Moscow State University

Title of the Russian original edition:
Differentsialnye uravneniya
Publisher Nauka, Moscow 1980

This volume is part of the *Springer Series in Soviet Mathematics*
Advisers: L. D. Faddeev (Leningrad), R. V. Gamkrelidze (Moscow)

ISBN-13:978-3-540-13002-4   e-ISBN-13:978-3-642-82175-2
DOI: 10.1007/978-3-642-82175-2

Library of Congress Cataloging in Publication Data
Tikhonov, A. N. (Andrei Nikolaevich), 1906–
Differential equations. (Springer series in Soviet Mathematics)
Translation of: Differentsial'nye uravneniia.
Bibliography: p. Includes index.
1. Differential equations. I. Vasil'eva, A. B. (Adelaida Borisovna), 1926–.
II. Sveshnikov, A. G. (Aleksei Georgievich), 1924–. III. Title. IV. Series.
QA371.T47413 1984   515.3'5   84-1444
ISBN-13:978-3-540-13002-4 (U.S.)

2141/3130-543210

Foreword

The proposed book is one of a series called "A Course of Higher Mathematics and Mathematical Physics" edited by A. N. Tikhonov, V. A. Ilyin and A. G. Sveshnikov.

The book is based on a lecture course which, for a number of years now has been taught at the Physics Department and the Department of Computational Mathematics and Cybernetics of Moscow State University. The exposition reflects the present state of the theory of differential equations, as far as it is required by future specialists in physics and applied mathematics, and is at the same time elementary enough.

An important part of the book is devoted to approximation methods for the solution and study of differential equations, e.g. numerical and asymptotic methods, which at the present time play an essential role in the study of mathematical models of physical phenomena. Less attention is paid to the integration of differential equations in elementary functions than to the study of algorithms on which numerical solution methods of differential equations for computers are based.

The reader will become acquainted with various methods for the numerical solution of initial values as well as boundary value problems, and with such fundamental notions of the theory of numerical methods as the convergence of difference schemes, approximation and stability. The chapter concerned with asymptotic methods contains, in particular, information on the so-called method of singular perturbations (the averaging method), the method of boundary functions, the WKB method; these methods have rapidly developed in the last decade in connection with the requirements of such branches of physics and technology as the theory of automatic control, hydrodynamics, quantum mechanics, kinetics, the theory of non-linear oscillations, etc.

The English translation of the book includes important changes from the first Russian edition. The most important ones concern the existence and uniqueness theorem for the solution of initial value problems. The new proofs are based on the method of differential inequalities. These same ideas are applied in the study of the dependence on parameters of solutions of differential equation systems. The use of the method of differential inequalities considerably simplifies the proofs, makes them more uniform and allows us to state the results in more general form.

The manuscript of the book was read through by E. A. Grebennikov and L. D. Kudryavtsev, who made a number of important remarks. Inestimable assistance in the preparation of the manuscript for publication was rendered by B. I. Volkov. To all of them the authors express their sincere gratitude.

The authors

Contents

Chapter I
Introduction

§1. The Concept of a Differential Equation

The present book is concerned with differential equations, i. e. relations between an unknown function, its derivatives and independent variables. Equations containing derivatives with respect to several independent variables are called *partial differential equations*. Equations containing derivatives with respect to only one of the independent variables are called *ordinary differential equations*. This book mainly deals with the properties and solution methods of ordinary differential equations; only the last chapter is devoted to certain special classes of partial differential equations.

The independent variable with respect to which the derivatives in an ordinary differential equation are taken is usually denoted by the letter x (or the letter t, since time often plays the role of the independent variable). The unknown function is denoted by $y(x)$.

An ordinary differential equation may be written as a relation of the form

$$F\left(x, y, \frac{dy}{dx}, \ldots, \frac{d^n y}{dx^n}\right) = 0. \tag{1.1}$$

Besides the unknown function, its derivatives with respect to the independent variable x and the independent variable x itself, equation (1.1) may involve additional variables $\mu_1, \ldots, \mu_k$. In this case we say that the unknown function depends on the variables $\mu_1, \ldots, \mu_k$ as parameters.

The order of the highest order derivative contained in equation (1.1) is known as the *order of the equation*. A first order equation is of the form

$$F\left(x, y, \frac{dy}{dx}\right) = 0 \tag{1.2}$$

and is a relation between three expressions – the unknown function, its derivative and the independent variable. It is often possible to write this relation in the form

$$\frac{dy}{dx} = f(x, y). \tag{1.3}$$

Equation (1.3) is called a first order equation *resolved* with respect to the derivative. We shall begin our study of the theory of ordinary differential equations with equation (1.3).

Along with differential equations (1.1)–(1.3) with one unknown function, the theory of ordinary differential equations deals with systems of equations. A system of first order equations resolved with respect to the derivatives

$$\frac{dy_i}{dx} = f_i(x, y_1, \ldots, y_n), \qquad (i = 1, \ldots, n), \tag{1.4}$$

is called a *normal system*. Introducing vector functions

$$y = (y_1, \ldots, y_n), \quad f = (f_1, \ldots, f_n),$$

we may write the system (1.4) in vector form

$$\frac{dy}{dx} = f(x, y). \tag{1.5}$$

It is easy to see that the *n*-th order equation (1.1), resolved with respect to the highest order derivative

$$\frac{d^n y}{dx^n} = f\left(x, y, \frac{dy}{dx}, \ldots, \frac{d^{n-1} y}{dx^{n-1}}\right), \tag{1.6}$$

may be reduced to a normal system. Indeed, introducing the notation

$$y(x) = y_1(x), \quad \frac{dy}{dx} = \frac{dy_1}{dx} = y_2(x), \ldots, \frac{d^{n-1} y}{dx^{n-1}} = \frac{dy_{n-1}}{dx} = y_n(x). \tag{1.7}$$

and using the obvious equality

$$\frac{d^n y}{dx^n} = \frac{dy_n}{dx},$$

we obtain the normal system

$$\frac{dy_1}{dx} = y_2,$$

$$\ldots\ldots\ldots$$

$$\frac{dy_{n-1}}{dx} = y_n, \tag{1.8}$$

$$\frac{dy_n}{dx} = f(x, y_1, \ldots, y_n).$$

corresponding to equation (1.6).

In equations (1.1)–(1.5), the independent variable will be assumed real. The unknown functions may be real-valued as well as complex-valued functions of a real variable.

Obviously, if

$$y(x) = \bar{y}(x) + i\bar{\bar{y}}(x),$$

where $\bar{y}(x)$ and $\bar{\bar{y}}(x)$ are the real and imaginary parts of the function $y(x)$ respectively, then equation (1.3) is equivalent to the following system of ordinary differential equations for real-valued functions

$$\frac{d\bar{y}}{dx} = \text{Re } f(x, \bar{y}, \bar{\bar{y}}), \qquad \frac{d\bar{\bar{y}}}{dx} = \text{Im } f(x, \bar{y}, \bar{\bar{y}}).$$

The solution of the system of differential equations (1.4) is, by definition, any family of functions which satisfy the equations identically. As a rule, and this will be clear from examples given below (see §2), if a differential equation is soluble, then it has an infinite set of solutions. The procedure of finding the solutions is known as *integration of differential equations*.

Any solution $y_i(x)$ $(i = 1, \ldots, n)$ of the system (1.4) may be interpreted geometrically as a curve in the $(n+1)$-dimensional space of the variables $x, y_1, \ldots, y_n$; this curve is called the *integral curve*. The subspace of the variables $y_1, \ldots, y_n$ is called the *phase space*, while the projection of the integral curve on the phase space is the *phase trajectory*.

Equation (1.4) determines a direction given by the vector $\tau = (1, f_1, \ldots, f_n)$ at every point of the domain D. Such a domain in space, with a direction given at each point, is said to be a *direction field*. The integration of the system of equations (1.4) may be interpreted geometrically as finding curves whose tangents at each point coincide with the direction τ determined by the given field of directions at that point.

As we pointed out above, differential equations have an infinite set of solutions in general. Therefore, when we integrate the system (1.4), we will find an infinite set of integral curves contained in the domain where the right-hand sides of the system (1.4) are defined. In order to distinguish an individual integral curve in the set of all solutions, thus specifying a so-called *particular solution* of the system (1.4), we must impose additional conditions. In many cases such additional conditions are the *initial conditions*

$$y_i(x_0) = y_i^0 \qquad (i = 1, \ldots, n), \tag{1.9}$$

which determine a point of $(n+1)$-dimensional space of the variables $x, y_1, \ldots, y_n$ through which the required integral curve passes.

The problem of integrating system (1.4) with initial conditions (1.9) is known as the *initial value problem* or *Cauchy problem*.

In the simplest case of one equation

$$\frac{dy}{dx} = f(x, y), \tag{1.10}$$

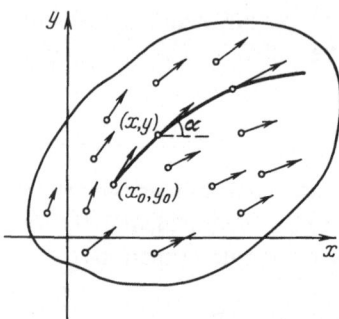

Fig. 1

the function $f(x, y)$ determines a direction field in the domain D (of the (x, y)-plane) where the right-hand side of (1.10) is defined. This direction field is given at each point of the domain D by the vector $\tau(x, y)$ with slope $f(x, y)$ ($\tan \alpha = f(x, y)$) (Fig. 1).

In order to solve the initial value problem with the condition $y(x_0) = y_0$ in this case, we must construct, in the domain D, the integral curve $y = y(x)$ which starts at the initial point (x_0, y_0) and is tangent to the vector τ of slope $f(x, y)$ at every one of its points (x, y).

Theorem 1.1 (*Chaplygin's theorem on differential inequalities*). *If for* $x \in [x_0, X]$ *there exists a solution of the initial value problem*

$$\frac{dy}{dx} = f(x, y), \ y(x_0) = y_0, \tag{1.11}$$

and, if $z(x)$ is a continuous and continuously differentiable function on $[x_0, X]$ satisfying

$$\frac{dz}{dx} < f(x, z), \ x \in [x_0, X],$$

$$z(x_0) < y_0, \tag{1.12}$$

then we have the inequality

$$z(x) < y(x), \ x \in (x_0, X]. \tag{1.13}$$

Indeed, by the assumptions of the theorem, inequality (1.13) is satisfied at the point x_0. Therefore, by the continuity of $y(x)$ and $z(x)$, it is satisfied also in some neighbourhood to the right of the point x_0. Assume that $x_1 \in [x_0, X]$ is the nearest point to x_0 in which the inequality (1.13) fails to hold, i. e. $z(x_1) = y(x_1)$. Geometrically, this means that the curves $z(x)$ and $y(x)$ intersect or are tangent when $x = x_1$. But then we must have $\frac{dz}{dx}(x_1) \geq f(x_1, y(x_1))$, which contradicts (1.12). The theorem is proved.

Now for some remarks – which will be important later – about the theorem on differential inequalities just proved above.

Remarks. 1. We assumed that $z(x_0) < y_0$, but the theorem remains valid if $z(x_0) = y_0$. In this case, the existence of a neighbourhood (to the right of the point x_0) in which the inequality (1.13) holds follows from the fact that

$$\frac{dz}{dx}(x_0) < f(x_0, z(x_0)) = f(x_0, y_0) = \frac{dy}{dz}(x_0)$$

The rest of the argument is exactly the same as in the case $z(x_0) < y_0$

2. If the function $z(x)$ satisfies the inequalities

$$\frac{dz}{dx} \geq f(x, z), \ x \in [x_0, X]$$

$$z(x_0) = y_0,$$

then the sign of the inequality in (1.13) should also be changed to the opposite one.

3. The theorem remains valid in the case when $z(x)$ is piecewise differentiable on $[x_0, X]$ and the inequality (1.12) is satisfied for the limiting values of the derivative $\frac{dz}{dx}$ at the points of discontinuity.

In order to answer a number of questions, it is convenient to reduce certain problems concerning differential equations to corresponding problems about integral equations.

Lemma 1.1. *Suppose $f(x, y)$ is a continuous function of the point (x, y) in some rectangle $D = \{|x - x_0| \leq a, |y - y_0| \leq b\}$. Then the initial value problem*

$$\frac{dy}{dx} = f(x, y), \ y(x_0) = y_0 \tag{1.14}$$

is equivalent to the integral equation

$$y(x) = y_0 + \int_{x_0}^{x} f(\xi, y(\xi)) \, d\xi. \tag{1.15}$$

Proof. Suppose that there exists a solution $y(x)$ of the initial value problem on the segment $|x - x_0| \leq a$ and we have $y_0 - b \leq y(x) \leq y_0 + b$ (these inequalities mean that for $|x - x_0| \leq a$ the integral curve is located within the domain D where $f(x, y)$ is continuous). Substituting $y(x)$ into equation (1.14), we obtain an identity. Integrating this identity from x_0 to $x \in [x_0 - a, x_0 + a]$ and using the initial condition $y(x_0) = y_0$, we obtain (1.15). Therefore, the solution of the initial value problem (1.14) satisfies the integral equation (1.15). On the other hand, if there exists a continuous solution of the integral equation (1.15) – the function $y(x)$, where $y_0 - b \leq y(x) \leq y_0 + b$, then, by the continuity of $f(x, y)$ and the continuity of the function $f(\xi, y(\xi))$ as a function of ξ which follows, the integral in the right-hand side of (1.15) is a continuously differentiable function of ξ. Therefore, the left-hand side of (1.15), i.e. the function $y(x)$, possesses a continuous derivative and this derivative equals $f(x, y(x))$, so that $y(x)$ is a solution of equation (1.14). The fact that the initial condition is satisfied can be checked directly. The lemma is proved.

Remark. A similar theorem on equivalence holds also for systems of differential equations, i.e. for the problem (1.4), (1.9).

One usually considers systems (1.4) whose right-hand sides are continuous in some domain D where the unknown functions y_i and the independent variable x vary. Obviously, in this case the solution will be a continuously differentiable function. However, in applications, one often meets with equations whose right-hand sides have discontinuities in the variable x (for example, in the description of instantaneously applied forces or concentrated forces), therefore these solutions will also possess discontinuous derivatives. It is then natural to consider, as solutions of (1.4), continuous functions $y_i(x)$ with piecewise continuous derivatives. In substituting them into the equation, they are to be differentiated everywhere except at points of discontinuity and points where derivatives do not exist. It is natural to call such solutions generalised solutions.

Lemma 1.1 remains valid in the case when the function $f(x, y)$ is a piecewise continuous function of the variable x. Then the integral equation (1.15) has a continuous solution $y(x)$ which is a piecewise differentiable function of x. This solution satisfies equation (1.14) in those intervals where the function $f(x, y)$ is continuous.

There are other ways of specifying supplementary conditions which determine a particular solution of the system (1.4). Among them let us note: the so-called boundary value problems, in which a particular solution is determined by certain supplementary conditions at some points of its domain of definition; the eigenvalue problem, which involves determining certain parameters appearing in the equation so that particular solutions exist and satisfy some supplementary requirements; and the problem of finding periodic solutions and a number of other specifications uniquely determining the required particular solution of the equation.

§ 2. Physical Problems Leading to Differential Equations

In this section we shall present some typical problems in physics and mechanics whose study, by means of mathematical models, leads to the investigation of differential equations.

1. Radioactive disintegration. The physical law which describes the process of radioactive disintegration states that the rate of disintegration is negative and proportional to the amount of non-disintegrated matter at the given moment of time. The coefficient of proportionality α, which is a constant characterising the given type of matter, does not depend on time and is known as the disintegration coefficient. The mathematical expression of the law of radioactive disintegration is

$$\frac{dm}{dt} = -\alpha m(t), \tag{1.16}$$

where $m(t)$ is the amount of non-disintegrated matter at the moment of time t. This relation is a first order differential equation resolved with respect to the derivative.

It is easy to check (by direct substitution) that there is solution of (1.16) of the form

$$m(t) = Ce^{-\alpha t}, \tag{1.17}$$

where C is an arbitrary constant, which may be determined from some supplementary condition, for example, from the initial condition $m(t_0) = m_0$ specifying the amount of matter at the initial moment t_0. A particular solution of the corresponding initial value problem is

$$m(t) = m_0 e^{-\alpha(t-t_0)}. \tag{1.18}$$

One of the important physical characteristics of the process of radioactive disintegration is the half-life, the time T needed for the amount of non-disintegrated matter to decrease by half. It follows from (1.18) that

$$\frac{m_0}{2} = m_0 e^{-\alpha T},$$

so that we get

$$T = \frac{1}{\alpha} \log 2 \tag{1.19}$$

Note that equation (1.16) is the mathematical model not only of the process of radioactive disintegration, but also of many other processes of splitting or multiplication characterized by the fact that the rate of splitting (multiplication) is proportional to the amount of matter at the given moment of time, the coefficient of proportionality being a certain constant for the given process. As we shall see, a typical way of setting the problem for this class of equations is the initial value problem (the Cauchy problem).

2. The motion of a system of particles.   The mathematical models of motion for a system of particles of mass $m_i (i = 1, \ldots, N)$, usually accepted in theoretical mechanics, are the equations of motion which follow from Newton's second law:

$$m_i \frac{d^2 r_i}{dt^2} = F_i\left(t, r_j, \frac{dr_j}{dt}\right) \qquad (i,j=1,\ldots,N). \tag{1.20}$$

Here m_i are the particle masses, which do not change in time, r_i are the radius-vectors of the particles, and F_i is the force vector acting on the i-th particle and depending, in general, on time, on the coordinates of the i-th particle and the position of the particles of the system, as well as their velocities. The system (1.20) is a system of N vector equations of the second order. If the masses of the particles do not change in the process of motion, then, by denoting the Cartesian coordinates of the radius-vector r_i by x_i, y_i, z_i and introducing new variables $v_{ix} = \frac{dx_i}{dt}$, $v_{iy} = \frac{dy_i}{dt}$, $v_{iz} = \frac{dz_i}{dt}$ (the components of the velocity vector of the i-th particle), we can write (1.20) in the form of a normal system of first order equations

$$\frac{dx_i}{dt} = v_{ix}, \quad \frac{dy_i}{dt} = v_{iy}, \quad \frac{dz_i}{dt} = v_{iz},$$

$$\frac{dv_{ix}}{dt} = \frac{1}{m_i} F_{ix}, \quad \frac{dv_{iy}}{dt} = \frac{1}{m_i} F_{iy}, \quad \frac{dv_{iz}}{dt} = \frac{1}{m_i} F_{iz}. \tag{1.21}$$

The difficulty in integrating system (1.21) is mainly determined by the form of the right-hand sides, i.e. the functional dependence of the components of the force vectors on the variables $t, x_i, y_i, z_i, v_{ix}, v_{iy}, v_{iz}$. In many cases it is possible to obtain the value of a particular solution of the system with a given degree of precision only by means of numerical methods, using computers. A typical problem for the system (1.21) is the initial value problem, which consists in determining the trajectory of the particles when their position and velocity at the initial moment of time t_0 are given,

$$r_i(t_0) = r_i^0, \ v_i(t_0) = v_i^0, \tag{1.22}$$

the right-hand sides being given functions (given external forces acting on the system and interaction forces between the particles themselves). Another typical problem for the system (1.21) is the boundary value problem which consists in determining the trajectory passing through the given initial and terminal points in the phase space. This is the problem that must be solved when we calculate the trajectory of a spacecraft leaving the Earth for the Moon or for some planet.

In a number of cases, other means of specifying a particular solution of the system (1.21) are considered.

An important special case of the system (1.20) is the oscillation equation of the physical pendulum. Usually, by a physical pendulum one means an absolutely rigid body, which can rotate, under the action of the force of gravity,

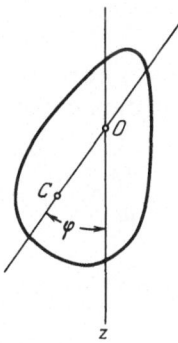

Fig. 2

about a motionless horizontal axis not passing through the centre of mass C (Fig. 2).

Consider a section of the solid by a plane perpendicular to the rotation axis and passing through the centre of mass. Denote the intersection point of the axis and the plane by O. Obviously, the position of the physical pendulum at any moment of time may be characterized by the angle φ made by the line OC with the vertical axis z passing through the point O. To deduce the equation of motion, let us use Newton's second law, as applied to rotational motion (the angular acceleration is proportional to the principal moment of the exterior forces). Then, ignoring friction forces, we obtain

$$I\frac{d^2\varphi}{dt^2} = -mgd \sin \varphi, \tag{1.23}$$

where I is the moment of inertia of the solid with respect to the axis of rotation and d is the distance from the point O to the centre of mass C.

The general equation (1.23) of oscillation of the physical pendulum is non-linear. In the case of small oscillations, restricting ourselves to the first term of the expansion of the function $\sin\varphi$, we get

$$\frac{d^2\varphi}{dt^2} + \omega^2\varphi = 0, \tag{1.24}$$

where ω^2 denotes the quotient $\omega^2 = mgd/I$. Obviously, from the point of view of dimension, $[\omega] = \sec^{-1}$, which justifies this notation. Note that in the case of equation (1.24) the returning force is proportional to the displacement from the position of equilibrium.

It is easy to check (by direct substitution) that equation (1.24) possesses periodic solutions of frequency ω

$$\varphi(t) = A \cos \omega t + B \sin \omega t, \tag{1.25}$$

where A and B are arbitrary constants determining the amplitude of periodic oscillations.

If we take into account friction forces proportional to angular velocity, equation (1.24) will become an equation of the form

$$\frac{d^2\varphi}{dt^2} + \alpha\frac{d\varphi}{dt} + \omega^2\varphi = 0. \tag{1.26}$$

As will be shown later (see Chapter 3) equation (1.26) determines damped oscillations.

3. The transfer equations. Suppose that air flows along a pipe of constant perpendicular section, whose axis coincides with the x axis, the velocity along the axis of the pipe at the point x at time t being a given function $v(x, t)$. Suppose the air carries a certain amount of matter whose linear density in the section of the pipe with coordinate x at time t will be denoted by $u(x, t)$. In the transfer process, some matter settles on the inner walls of the pipe. We shall assume that the density of distribution of the matter which settles is given by the expression $f(x,t)\ u(x,t)$ ($f(x,t)$ is a given function), i.e. is proportional to the concentration of matter; this may be viewed as a linear approximation (to a more complicated law) valid for sufficiently small u. This means that the amount of matter settling on the part of the pipe located between the sections x and $x + \Delta x$ during time $[t, t + \Delta t]$ is given by the expression

$$\int_x^{x+\Delta x} \int_t^{t+\Delta t} f(\xi, \tau)\,u(\xi, \tau)\,d\xi d\tau.$$

To obtain a differential equation with respect to u, consider the balance of matter in the domain between the sections x and $x + \Delta x$. The process of diffusion will not be taken into consideration, which is natural if the velocity v is sufficiently large.

During the period of time Δt, the change in the amount of matter in the domain under consideration equals

$$\int_x^{x+\Delta x} [u(\xi, t + \Delta t) - u(\xi, t)]\,d\xi.$$

This change is determined, firstly, by the difference of flows of matter: the amount which flows in through the section x and equals

$$\int_t^{t+\Delta t} v(x, \tau)\,u(x, \tau)\,d\tau$$

and the amount which flows out through the section $x + \Delta x$ and equals

$$\int\limits_{t}^{t+\Delta t} v(x+\Delta x,\tau)\, u(x+\Delta x,\tau)\, d\tau,$$

and, secondly, by the decrease of the amount of matter due to settling on the walls of the pipe, which equals

$$-\int\limits_{x}^{x+\Delta x}\int\limits_{t}^{t+\Delta t} f(\xi,\tau)\, u(\xi,\tau)\, d\xi d\tau.$$

Thus the law of conservation of matter gives

$$\int\limits_{x}^{x+\Delta x} [u(\xi,t+\Delta t)-u(\xi,t)]\, d\xi = \int\limits_{t}^{t+\Delta t} [v(x,\tau)\, u(x,\tau)-v(x+\Delta x,\tau)u(x+\Delta x,\tau)]$$

$$\cdot d\tau - \int\limits_{x}^{x+\Delta x}\int\limits_{t}^{t+\Delta t} f(\xi,\tau)\, u(\xi,\tau)\, d\xi d\tau. \tag{1.27}$$

Using the mean-value theorem of the differential calculus for the expressions under the integral signs, assuming that continuous partial derivatives of the given functions exist, and calculating the integrals by the mean value theorem of the integral calculus, we obtain

$$\frac{\partial u}{\partial t}(x^*,t)\big|_{t=t^*}\Delta x\Delta t = -\frac{\partial}{\partial x}(v(x,t^{**})u(x,t^{**})\big|_{x=x^{**}}\Delta x\Delta t$$

$$-f(x^{***},t^{***})\, u(x^{***},t^{***})\,\Delta x\Delta t, \tag{1.28}$$

where $x^*, x^{**}, x^{***}, t^*, t^{**}, t^{***}$ are certain points on the segments $[x, x+\Delta x]$, $[t, t+\Delta t]$ respectively. Dividing the relation (1.28) by $\Delta x\Delta t$ and assuming that Δx and Δt tend to zero, in view of the continuity of all terms of relation (1.28), we finally obtain the equation

$$\frac{\partial u}{\partial t}+\frac{\partial}{\partial x}(uv)+fu=0, \tag{1.29}$$

or

$$\frac{\partial u}{\partial t}+v(x,t)\frac{\partial u}{\partial x}+c(x,t)\, u=0, \tag{1.30}$$

where

$$c(x,t)=\frac{\partial v}{\partial x}(x,t)+f(x,t). \tag{1.31}$$

Equation (1.30) is a first order partial differential equation. The following problem, for example, may be considered for this equation. Suppose we know the concentration of matter for $x = x_0$

$$u(x_0, t) = u_0(t), \tag{1.32}$$

where $u_0(t)$ is a given function. We are to determine $u(x, t)$ when $x \geq x_0$. Further (Chapter 8) we shall see that condition (1.32) uniquely determines the solution of equation (1.30).

4. The percolation problem (water seeping through sand). Suppose water is seeping through sand vertically downwards. Direct the x-axis vertically down. By $u(x, t)$ denote the density of water in sand (t is time). The velocity of the water, v, obviously depends on its density, i.e. $v = v(u)$, where $v(u)$ is a given function, and v increases with u.

Consider the balance of water in the layer $[x, x + \Delta x]$. During time Δt, the change in the amount of water equals $\int\limits_{x}^{x + \Delta x} [u(\xi, t + \Delta t) - u(\xi, t)] \, d\xi$. This change is due to the difference between the input flow

$$\int\limits_{t}^{t + \Delta t} v(u(x, \tau)) \, u(x, \tau) \, d\tau$$

and the output flow

$$\int\limits_{t}^{t + \Delta t} v(u(x + \Delta x, \tau)) \, u(x + \Delta x, \tau) \, d\tau.$$

Thus

$$\int\limits_{x}^{x + \Delta x} [u(\xi, t + \Delta t) - u(\xi, t)] \, d\xi$$

$$= \int\limits_{t}^{t + \Delta t} [v(u(x, \tau)) \, u(x, \tau) - v(u(x + \Delta x, \tau)) \, u(x + \Delta x, \tau)] \, d\tau.$$

Assuming the existence of continuous partial derivatives of the function u and the differentiability of $v(u)$, apply the finite increment theorem and mean value theorem to compute the integral. Then divide by $\Delta x \Delta t$ and assume that Δx and Δt tend to zero, obtaining the equation

$$\frac{\partial u}{\partial t} + \frac{\partial u}{\partial x} \frac{d}{du} (v(u) u) = 0,$$

or

$$\frac{\partial u}{\partial t} + p(u) \frac{\partial u}{\partial x} = 0, \tag{1.33}$$

where

$$p(u) = v(u) + u \frac{dv}{du} \tag{1.34}$$

is a given function of u.

Equation (1.33), like Eq. (1.30), is a first order partial differential equation. Typical problems for it are obtained either by fixing the function $u(x, t)$ at a given value $x = x_0$:

$$u(x_0, t) = u_0(t)$$

(i. e. the density of water on the boundary of the layer of sand at all moments of time is given) or by fixing the function $u(x, t)$ at a given moment of time $t = t_0$

$$u(x, t_0) = u_1(x)$$

(i. e. by fixing the distribution of density of water along the entire section of sand at a definite moment of time t_0).

Note that Eq. (1.33), in which the factor $p(u)$ (which multiplies the derivative) depends on the unknown function, is more complicated than (1.30), in which both the derivative of the unknown function and the unknown function itself appear linearly. Equation (1.33) is known as a quasilinear equation. Quasilinear equations will be studied in Chapter 8.

5. Oscillations of an elastic rod. Consider the problem of small longitudinal oscillations of an elastic rod. Suppose that in the non-deformed state the rod is of length l, its axis coincides with the x-axis and, when oscillating under the action of external forces directed along the x-axis, the lateral sections of the rod move without deformation as rigid planes. Then the rod's oscillations may be characterized by one scalar function $u(x, t)$ – the value of the displacement at the moment of time t of the rod's section which, in its non-deformed state, has the coordinate x. We shall consider a rod of variable density $\varrho(x)$ statisfying Hooke's law: the elastic force deforming an infinitely small segment of the rod contained between sections x and $x + \Delta x$ is proportional to the relative length increase of this element. The coefficient of proportionality (known as the elasticity coefficient) will also be assumed variable along the rod and will be denoted by $k(x)$. Let us compute the relative increase of length ε of the element under consideration. Obviously, the length of this element at the moment of time t equals

$$\Delta l = (x + \Delta x) + u(x + \Delta x, t) - x - u(x, t) = \Delta x + u(x + \Delta x, t) - u(x, t),$$

so that for the relative increase of length we get

$$\varepsilon = \frac{\Delta l - \Delta x}{\Delta x} = \frac{u(x + \Delta x, t) - u(x, t)}{\Delta x}. \tag{1.35}$$

Passing to the limit as $\Delta x \to 0$ in the expression (1.35), assuming the function $u(x, t)$ to be continuously differentiable and using Hooke's law, we see that the force of elastic tension applied to the section x and acting on the left-hand side from the right-hand side equals $k(x)u_x(x, t)$. Note that the expression obtained for the force of elastic tension is valid only in the case of small oscillations (when it is possible to apply Hooke's law to an infinitely small element of the rod).

In order to obtain the equation of the oscillations of the rod, let us apply Newton's second law to the element considered. We shall assume that the external forces applied to the rod are distributed with density $f(x, t)$, so that the impulse of the force acting on the element Δx during the time interval Δt equals

$$\Delta I = \int_t^{t+\Delta t} \int_x^{x+\Delta x} f(\xi, \tau)\, d\xi\, d\tau. \tag{1.36}$$

Also, the forces of elastic tension defined above act on the boundary sections of the element considered. Then Newton's second law can be written in the form

$$\int_x^{x+\Delta x} [\varrho(\xi)\, u_t(\xi, t+\Delta t) - \varrho(\xi)\, u_t(\xi, t)]\, d\xi$$

$$= \int_t^{t+\Delta t} [k(x+\Delta x)\, u_x(x+\Delta x, \tau) - k(x)\, u_x(x, \tau)]\, d\tau + \int_t^{t+\Delta t} \int_x^{x+\Delta x} f(\xi, \tau)\, d\xi\, d\tau. \tag{1.37}$$

This is the integro-differential equation of oscillations of an elastic rod. Assuming that the functions in the square brackets under the integral expressions in (1.37) are continuously differentiable and that the function $f(x, t)$ is continuous, we can use the theorem on finite increments and compute the integrals by using the mean value theorem, thus obtaining

$$\varrho(x^*)\, u_{tt}(x^*, t)\big|_{t=t^*}\, \Delta x \Delta t$$

$$= \frac{\partial}{\partial x}\, (k(x)\, u_x(x, t^{**}))\big|_{x=x^{**}}\, \Delta x \Delta t + f(x^{***}, t^{***})\, \Delta x \Delta t,$$

where $x^*, x^{**}, x^{***}, t^*, t^{**}, t^{***}$ are certain points from the segments $[x, x+\Delta x]$, $[t, t+\Delta t]$ respectively. Dividing by $\Delta x \Delta t$ and passing to the limit as $\Delta x \to 0$, $\Delta t \to 0$, we obtain, in view of the smoothness conditions on the functions $u(x, t)$, $\varrho(x)$, $k(x)$, $f(x, t)$, the differential equation of longitudinal oscillations of an elastic rod

$$\varrho(x)u_{tt}(x, t) = [k(x)u_x(x, t)]_x + f(x, t). \tag{1.38}$$

This is a second order partial differential equation, which is the mathematical model of the oscillations in space and time of a continuous elastic medium. In the static case, the rod, under the action of an external force (which is constant in time) and of the forces of elastic interaction, assumes a certain state of static equilibrium, which is described by the ordinary differential equation

$$[k(x)u_x(x)]_x + f(x) = 0. \tag{1.39}$$

A typical problem for equation (1.39) is the boundary value problem which arises when the displacements of the boundary points of the rod are given

$$u(0)=u_0, \; u(l)=u_1, \tag{1.40}$$

or the tension applied to the boundary section is given

$$k(0)u_x(0)=f_1, \; k(l)u_x(l)= -f_2. \tag{1.41}$$

In a number of cases one considers other boundary value problems as well.

The general equation of oscillations of a distributed system (1.38) also becomes an ordinary differential equation in the case of periodic oscillations taking place under the action of a periodic external force. Suppose $f(x,t)=\tilde{f}(x)\cos\omega t$. We shall look for the solution of (1.38) in the form $u(x,t)=\tilde{u}(x)\cos\omega t$. Then $\tilde{u}(x)$, the amplitude of periodic oscillations established in the system under the action of the periodic external force, will be given by the ordinary differential equation

$$[k(x)\tilde{u}_x(x)]_x+\omega^2\varrho(x)\tilde{u}(x)= -\tilde{f}(x). \tag{1.42}$$

A typical boundary value problem specifying a particular solution of equation (1.42) is again the boundary value problem with boundary conditions of the type (1.40), (1.41) or of more complex form.

In a number of cases we are interested in determining the eigen-frequencies of oscillation of the system, i. e. the frequencies of established periodic oscillations which are possible in the system in the absence of external forces. This problem reduces to the boundary value problem for the homogeneous equation (1.42):

$$[k(x)u_x(x)]_x+\omega^2\varrho(x)u(x)=0. \tag{1.43}$$

It is required to determine the values of the parameter ω^2 for which equation (1.43) possesses a non-trivial solution, statisfying given homogeneous boundary conditions. Such a problem is known as a boundary value problem on eigenvalues.

6. The equation of heat conduction. One of the typical equations of mathematical physics is the heat equation, which we shall now derive. The caloric state of the solid D may be described by means of a scalar function $u(M,t)$, the temperature at the point M of the solid at the moment of time t. The physical characteristics of the body are described by functions of density $\varrho(M)$ and heat capacity $c(M)$ which, in a wide range of temperatures, may be assumed to be independent of temperature, as well as by the coefficient of heat conduction $k(M)$, which is the coefficient of proportionality between the density of heat flow through an elementary surface ΔS and the temperature gradient in the direction normal to this surface

$$\Delta q = -k(M) \frac{\partial u}{\partial n} \Delta S. \qquad (1.44)$$

We assume that the flow of heat is directed from the higher temperature side of the surface to the lower temperature side (the temperature gradient in this direction is negative); this explains the minus sign in formula (1.44).

In order to construct a mathematical model of temperature change in the solid under consideration, let us write out the balance equation. The change of the amount of heat in the volume element ΔV during the interval of time from t to $t + \Delta t$ is

$$\Delta Q_1 = \int_{\Delta V} c(M) \varrho(M) \, [u(M, t + \Delta t) - u(M, t)] dV \qquad (1.45)$$

and is determined by the flow of heat through the boundary surface $\Delta \Sigma$ of the volume element

$$\Delta Q_2 = \int_t^{t+\Delta t} \int_{\Delta \Sigma} k(M) \frac{\partial u}{\partial n} \, d\sigma d\tau \qquad (1.46)$$

(the derivative is taken along the direction of the exterior normal, which explains the plus sign in front of the integral in (1.46)) and by the amount of heat supplied by the external sources distributed in space and in time with density $f(M, t)$:

$$\Delta Q_3 = \int_t^{t+\Delta t} \int_{\Delta V} f(M, \tau) dV d\tau. \qquad (1.47)$$

We have

$$\Delta Q_1 = \Delta Q_2 + \Delta Q_3. \qquad (1.48)$$

Transforming the surface integral in the expression for ΔQ_2 according to Ostrogradsky's formula (here we assume the neccessary smoothness of functions $k(M)$ and $u(M, t)$), we obtain an integral relation for the balance of heat in the form

$$\int_{\Delta V} c(M) \varrho(M) \, [u(M, t + \Delta t) - u(M, t)] dV$$

$$= \int_t^{t+\Delta t} \int_{\Delta V} \mathrm{div} \, [k(M) \, \mathrm{grad} \, u(M, \tau)] \, dV d\tau + \int_t^{t+\Delta t} \int_{\Delta V} f(M, \tau) dV d\tau. \quad (1.49)$$

Replacing the expression in the square brackets on the left hand side of (1.49) according to the finite increment theorem, we can calculate the integral by using the mean value theorem and pass to the limit as $\Delta V \rightarrow 0$, $\Delta t \rightarrow 0$ in the expression obtained; thus we get a differential equation satisfied by the temperature inside the solid D:

$$c(M)\varrho(M)\frac{\partial u}{\partial t}=\text{div } [k(M)\text{ grad }u(M,t)]+f(M,t). \qquad (1.50)$$

Here, just as in our derivation of the equation of elastic oscillations (1.38), we assume that the unknown function and the coefficients in the equation (1.50) possess the required number of derivatives.

Equation (1.50) is a partial differential equation – we have constructed a mathematical model of the change of temperature in space and time. A stationary distribution of temperature under the effect of sources which do not depend on time is described by the equation

$$\text{div } [k(M)\text{ grad }u(M)]+f(M)=0. \qquad (1.51)$$

This, in general, is also a partial differential equation – the temperature depends on certain coordinates in space. In the particular case of a stationary distribution of temperature depending on only one of the space coordinates, say in the case of temperature distributed along a rod with longitudinal axis x, we obtain an ordinary differential equation

$$[k(x)u_x(x)]_x+f(x)=0. \qquad (1.52)$$

Typical problems for specifying a particular solution of equation (1.52), as well as in the problem (1.39), are boundary value problems with given boundary conditions.

Chapter II
General Theory

§1. Elementary Integration Methods

The solution of a differential equation, as a rule, cannot be expressed in terms of elementary functions or in the form of quadratures of such functions; in order to obtain particular solutions, it is often necessary to appeal to different numerical methods; their efficiency has considerably increased with the appearance and development of modern computers. However, before the computer age, the desire to "integrate a differential equation in quadratures" determined one of the major directions in the study of ordinary differential equations and brought about the appearance of numerous reference books e. g. [8] on solutions of differential equations. In this section, we will briefly consider some of the simplest and most frequently applied cases in which the solution of the first order equation

$$\frac{dy}{dx} = f(x, y) \tag{2.1}$$

can be obtained in quadratures. In many cases neither of the variables x nor y in (2.1) plays a preferred role. This leads us to consider, along with equation (2.1), the equation

$$\frac{dx}{dy} = \frac{1}{f(x, y)}, \tag{2.2}$$

and also the first order equation written in the form

$$f_1(x, y)dx + f_2(x, y)dy = 0.$$

1. Equations with separable variables. These are equations of the form

$$\frac{dy}{dx} = \frac{f_1(x)}{f_2(y)} \tag{2.3}$$

or

$$f_1(x)dx + f_2(y)dy = 0. \tag{2.4}$$

Assume that the equation (2.4) has a solution in a certain domain D of the (x, y)-plane. Let

$$D = \{|x - x_0| \leq a, \, |y - y_0| \leq b\}.$$

The functions $f_1(x)$ and $f_2(y)$ are defined and continuous respectively on $|x - x_0| \leq a$ and $|y - y_0| \leq b$. Substituting this solution into (2.4), we obtain an identity and, after integration, we get

$$\int f_1(x)dx + \int f_2(y)dy = \text{const.} \tag{2.5}$$

The indefinite integrals in (2.5) are known as *quadratures*, and this is where the expression "integration of an equation in quadratures" comes from. The relation (2.5) may be rewritten in the form

$$\Phi(x, y) = C, \tag{2.6}$$

which shows that the function $\Phi(x, y)$ assumes constant values on the solutions of equation (2.4) (to different solutions correspond different values of the constant).

For every fixed value of C, the expression (2.6) determines a certain particular solution $y = y(x)$ of equation (2.4) as an implicit function of the variable x. But if we consider C as a parameter, then the expression (2.6) determines the family of solutions $y = y(x, C)$. The expression (2.6) is known as the *integral* of the corresponding differential equation. If the expression (2.6), or in a more general case an expression of the form $\Phi(x, y, C) = 0$, in which C is viewed as a parameter, determines the entire set of solutions of the corresponding differential equation, then this expression is known as the *general integral* of the given differential equation and any expression $y = y(x, C)$ obtained from it and containing all the solutions of the given equation is referred to as the *general solution* of the given differential equation.

The expression (2.5) is, obviously, the general integral of equation (2.4).

In order to specify the particular solution of equation (2.4) determined by the initial condition

$$y(x_0) = y_0, \tag{2.7}$$

it is sufficient to find the constant C_1 from the expression of the general integral (2.5), written in the form

$$\int_{x_0}^{x} f_1(x)dx + \int_{y_0}^{y} f_2(y)dy = C_1,$$

The initial condition requirement implies $C_1 = 0$; hence the required particular solution in implicit form is determined by the integral

$$\int_{x_0}^{x} f_1(x)dx + \int_{y_0}^{y} f_2(y)dy = 0. \tag{2.8}$$

It is easy to see that many equations may be transformed into equations with separable variables of the form (2.4). Thus the equation

$$f_1(x)g_1(y)dx + f_2(x)g_2(y)dy = 0,$$

after it is divided by $g_1(y)f_2(x)$, assumes the necessary form. One must keep in mind that in this case we may lose particular solutions for which the product $g_1(y)f_2(x)$ vanishes.

Example 2.1. The simplest equation with separable variables is the equation

$$\frac{dy}{dx} = f(x). \tag{2.9}$$

Its general integral is of the form

$$y - \int f(x)dx = C \tag{2.10}$$

and the particular solution satisfying the initial condition (2.7) is

$$y = \int_{x_0}^{x} f(x)dx + y_0. \tag{2.11}$$

In the particular case of equation (2.9) with constant right-hand side

$$\frac{dy}{dx} = a, \tag{2.12}$$

we obtain a particular solution which satisfies the initial condition (2.7) in the form

$$y = a(x - x_0) + y_0. \tag{2.13}$$

Example 2.2.
$$\frac{dy}{dx} = f\left(\frac{y}{x}\right). \tag{2.14}$$

Let us carry out a change of the unknown variable $z = y/x$. Since we then obtain $y = xz$, $\frac{dy}{dx} = x\frac{dz}{dx} + z$, Eq. (2.14) is transformed into the equation

$$x\frac{dz}{dx} + z = f(z),$$

which can now be written in the form (2.4):

$$\frac{dx}{x} = \frac{dz}{f(z) - z}.$$ (2.15)

The equation

$$\frac{dy}{dx} = f(ax + by),$$ (2.16)

where a and b are constants, can also be transformed into the form (2.4). Introduce new variable $z = ax + by$. Then

$$\frac{dz}{dx} = a + b\,\frac{dy}{dx}$$

and Eq. (2.16) becomes

$$\frac{dz}{dx} = a + b\,f(z)$$ (2.17)

or

$$\frac{dx}{dz} = \frac{1}{a + bf(z)},$$ (2.18)

the equation considered in example 2.1.

Now let us consider the equation

$$M(x, y)dx + N(x, y)dy = 0,$$ (2.19)

where $M(x, y)$ and $N(x, y)$ are homogeneous functions in the variables x and y of the same degree. A function $f(x, y)$ is said to be a homogeneous function of the variables x and y of degree k if we have the relation

$$f(tx, ty) = t^k f(x, y).$$ (2.20)

Note that $f\!\left(\dfrac{y}{x}\right)$ is a homogeneous function of degree zero.

Writing equation (2.19) in the form

$$\frac{dy}{dx} = -\frac{M(x, y)}{N(x, y)},$$ (2.21)

we see that under our assumptions concerning the functions $M(x, y)$ and $N(x, y)$ the right-hand side of (2.21) is a homogeneous function of degree zero and therefore, after the substitution $z = y/x$ (just as in example 2.2) Eq. (2.19) will become an equation with separable variables of the (2.4) type.

2. First order linear equations. An equation is said to be linear, if it is linear with respect to the unknown function and its derivatives. Linear equations of the first order are of the form

$$\frac{dy}{dx}+p(x)y(x)=f(x).\tag{2.22}$$

If $f(x)=0$, then this equation is called homogeneous. As can be easily seen, the linear homogeneous equation

$$\frac{dy}{dx}+p(x)\,y(x)=0\tag{2.23}$$

can be transformed into an equation with separable variables $\dfrac{dy}{y}+p(x)dx=0$ whose general integral will be of the form

$$\log|y|+\int p(x)dx=C_1,\tag{2.24}$$

and the general solution is

$$y=Ce^{-\int p(x)dx},\tag{2.25}$$

where $C\neq0$. Obviously, the particular solution $y(x)\equiv0$ of the equation (2.23), which we have lost by dividing (2.22) by y, is contained in formula (2.25) in the case $C=0$. Therefore (2.25), where C is now any real number, is the general solution of equation (2.23).

By replacing the indefinite integral $\int p(x)dx$ in (2.25) by the definite integral $\int_{x_0}^{x} p(x)dx$, we obtain a particular solution of equation (2.23) which satisfies the initial condition $y(x_0)=y_0$ in the form

$$y=y_0e^{-\int_{x_0}^{x} p(x)dx}.\tag{2.26}$$

Note that the actual method for constructing formula (2.26) is in fact a proof of the uniqueness of the solution of the initial value problem for Eq. (2.23) under the assumption that this solution exists. Indeed, by substituting any solution of the initial value problem in Eq. (2.23) and carrying out successively the transformations (2.24)–(2.26), we shall always obtain the same result, namely formula (2.26). In order to prove the existence of the solution of the given problem, it is sufficient to carry out a direct substitution in the equation, which shows that for any continuous function $p(x)$ the function $y(x)$ defined by formula (2.26) satisfies all the conditions of the initial value problem for Eq. (2.23). Obviously, arguments of this type may be carried out as well in the case of initial value problems for equations with separable variables considered in subsection 1 of the present section.

The solution of the linear non-homogeneous equation (2.22) will be found by the method of variation of parameters, which involves using a special substitution of the unknown function

$$y(x) = C(x)e^{-\int p(x)dx}, \tag{2.27}$$

where $C(x)$ is a function which must be determined. Substituting this form of the solution into the equation, we obtain

$$\frac{dC}{dx}e^{-\int p(x)dx} - C(x)p(x)e^{-\int p(x)dx} + p(x)C(x)e^{-\int p(x)dx} = f(x),$$

hence

$$\frac{dC}{dx} = f(x)e^{\int p(x)dx}.$$

Integrating, we get

$$C(x) = \int f(x)e^{\int p(x)dx}dx + C_1,$$

and finally

$$y(x) = C_1 e^{-\int p(x)dx} + e^{-\int p(x)dx}\int f(x)e^{\int p(x)dx}dx. \tag{2.28}$$

It follows from the expression obtained above that the general solution of the linear non-homogeneous equation (2.22) can be represented as the sum of the general solution (2.25) of the linear homogeneous equation (2.23) and a particular solution of the non-homogeneous equation (2.22); this can be easily checked by substituting the second summand of formula (2.28) in the non-homogeneous equation (2.22).

The solution of the initial value problem $y(x_0) = y_0$ for Eq. (2.22) will be found by determining the constant C_1 in formula (2.28) from the initial conditions. To do this, it is convenient to write the indefinite integrals in (2.28) in the form of definite integrals $\int_{x_0}^{x}$. Then $C_1 = y_0$ and we get

$$y(x) = y_0 e^{-\int_{x_0}^{x} p(\xi)d\xi} + e^{-\int_{x_0}^{x} p(\xi)d\xi}\int_{x_0}^{x} f(\xi)e^{\int_{x_0}^{\xi} p(\xi)d\xi}d\xi,$$

i.e.

$$y(x) = y_0 e^{-\int_{x_0}^{x} p(\xi)d\xi} + \int_{x_0}^{x} e^{-\int_{\xi}^{x} p(\xi)d\xi}f(\xi)d\xi. \tag{2.29}$$

Thus the required solution is determined as the sum of the solution of the homogeneous equation (2.23) satisfying the given initial condition $y(x_0) = y_0$ and the solution of the non-homogeneous equation satisfying the zero initial condition.

The representation (2.29) was obtained under the assumption that the solution exists. It proves the uniqueness of the solution of the initial value problem for the non-homogeneous equation (2.22).

The uniqueness of this solution may also be established by the following arguments, which are typical of linear problems. Assume that there exist two different solutions of the initial value problem $y_1(x)$ and $y_2(x)$. Consider their difference $z(x) = y_1(x) - y_2(x)$. Obviously the function $z(x)$ is a solution of the initial value problem for the corresponding homogeneous equation with zero initial condition

$$\frac{dz}{dx} + p(x)z(x) = 0, \quad z(x_0) = 0.$$

Hence, in view of the uniqueness of the solution of the initial value problem for the linear homogeneous equation, it follows that $z(x) \equiv 0$.

The existence of the initial value problem's solution for Eq. (2.22) in the case of continuous functions $p(x)$ and $f(x)$ can be established directly by substituting formula (2.29) into the equation and into the initial condition.

Now let us obtain an important estimate of the increase of the initial value problem's solution for the linear equation. Suppose that in Eq. (2.22) the functions $p(x)$ and $f(x)$ satisfy, on the interval of variation of the independent variable which we consider, the condition

$$|p(x)| \leq K, \quad |f(x)| \leq M. \tag{2.30}$$

Then from the representation (2.29) we have the following estimate for the solution of the initial value problem

$$|y(x)| \leq |y_0| e^{K |x - x_0|} + \frac{M}{K} (e^{K |x - x_0|} - 1). \tag{2.31}$$

To conclude this subsection, we shall mention some equations which often appear in applications and which, by an appropriate change of variables, may be reduced to linear equations. Consider the so-called Bernoulli equation

$$\frac{dy}{dx} + p(x)y = f(x)y^n,$$

where $n \neq 1$, otherwise the equation is already linear. Introduce the new unknown function $z = y^{1-n}$. Then the Bernoulli equation will become a linear equation

$$\frac{1}{1-n} \frac{dz}{dx} + p(x)z(x) = f(x),$$

whose general solution is given by formula (2.28).

The more complicated Riccati equation

$$\frac{dy}{dx}+p(x)y(x)+q(x)y^2(x)=f(x)$$

cannot be integrated in quadratures in the general case. However, it possesses the following important property: if we know at least one particular solution $y=y_1(x)$ of the Riccati equation, then the problem of finding its general solution reduces to solving a linear equation. Indeed, by introducing the new unknown function

$$z(x)=y(x)-y_1(x),$$

we obtain the Bernoulli equation

$$\frac{dz}{dx}+[p(x)+2q(x)\,y_1(x)]z(x)+q(x)z^2(x)=0$$

for z, proving the above-mentioned property.

3. The lemma on differential inequalities. Further we shall need the following statement.

Lemma 2.1 (*the lemma on differential inequalities*). *Suppose the function $z(x)$ is continuous and possesses a piecewise continuous derivative on $[x_0, X]$, statisfying the inequality*

$$\left|\frac{dz}{dx}\right|\leq N|z|+a, \tag{2.32}$$

where $N>0$, $a>0$ are positive constants. (At the discontinuity points of the derivative the inequality (2.32) is satisfied by its limiting values). Then we have the estimate

$$|z(x)|\leq s(x), \quad x\in(x_0, X], \tag{2.33}$$

where $s(x)$ is the solution of the initial value problem for the linear differential equation

$$\frac{ds}{dx}=Ns+a, \quad s(x_0)=b\geq|z(x_0)|. \tag{2.34}$$

The proof of this lemma is similar to that of the Chaplygin theorem. Consider the solution of the initial value problem $\dfrac{d\tilde{s}}{dx}=\tilde{N}\tilde{s}+a$, $\tilde{s}(x_0)=b$, where $\tilde{N}>N$ is an arbitrary positive number. Note that $\tilde{s}(x)\geq0$.

Since $\dfrac{dz}{dx}(x_0) < \dfrac{d\tilde{s}}{dx}(x_0)$, it follows that in some neighbourhood to the right of the point x_0 we have the inequality $z(x) < \tilde{s}(x)$. Suppose $x = x_1 > x_0$ is the smallest value of x for which this inequality breaks down, i.e. $z(x_1) = \tilde{s}(x_1)$. But then

$$\frac{dz}{dx}(x_1) \geq \frac{d\tilde{s}}{dx}(x_1) = \tilde{N}\tilde{s}(x_1) + a = \tilde{N}z(x_1) + a$$

which contradicts (2.32). (If x_1 is a point where the derivative is discontinuous, then, in the inequality, we must write the derivative of z from the left at the point x_1 in place of $\dfrac{dz}{dx}$). Similarly, from the inequality $\dfrac{dz}{dx}(x_0) > -\dfrac{d\tilde{s}}{dx}(x_0)$, it follows that we have the inequality $z(x) > -\tilde{s}(x)$ in a neighbourhood of x_0 and the first violation of this inequality for $x_2 > x_0$, i.e. the equality $z(x_2) = -\tilde{s}(x_2)$, yields $\dfrac{dz}{dx}(x_2) \leq -\dfrac{d\tilde{s}}{dx}(x_2)$, hence we again obtain a contradiction to 2.32).

Thus we have the estimate

$$|z(x)| < \tilde{s}(x) = be^{\tilde{N}(x - x_0)} + \frac{a}{\tilde{N}}(e^{\tilde{N}(x - x_0)} - 1)$$

(for $\tilde{s}(x)$ here we have written out the explicit form which can be found, for example, by using formula (2.29)). Since $\tilde{N}$ is an arbitrary number greater than $N(\tilde{N} = N + \varepsilon, \varepsilon > 0)$, if we pass to the limit in this inequality as $\varepsilon \to 0$, we obtain the estimate

$$|z(x)| \leq be^{N(x - x_0)} + \frac{a}{N}(e^{N(x - x_0)} - 1) = s(x),$$

where $s(x)$ is the solution of the initial value problem (2.34). The lemma is proved.

Let us write the function $s(x)$ in a form which will be convenient later:

$$s(x) = bs_1(x) + as_2(x), \tag{2.35}$$

where

$$s_1(x) = e^{N(x - x_0)} \tag{2.36}$$

and

$$s_2(x) = \frac{1}{N}(e^{N(x - x_0)} - 1). \tag{2.37}$$

Remarks. The estimate (2.31) can easily be obtained from the lemma which we have proved above.

§2. Theorems on the Existence and Uniqueness of the Solution of the Initial Value Problem for a First Order Equation Resolved with Respect to the Derivative. The Euler Polygonal Line Algorithm

In the previous section, using explicit formulas, we proved the existence and uniqueness of the initial value problem's solution for the linear equation (2.22). We now consider the corresponding theorems for the initial value problem

$$\frac{dy}{dx} = f(x, y), \quad y(x_0) = y_0 \tag{2.38}$$

under more general assumptions concerning the function $f(x, y)$. Here the proof will also be carried out constructively: together with the proof of the existence theorem for the solution of problem (2.38) we shall obtain an algorithm for constructing a function $\bar{y}(x)$ which approximates the solution of the given problem as precisely as we wish. The idea of this method comes from Euler. The method consists in replacing the integral curve (which is the solution of problem (2.38)) in successive steps by a certain polygonal line – the Euler polygonal line.

Suppose the function $f(x, y)$ is given in the domain G of the (x, y)-plane. In this domain, construct the closed rectangle

$$D = \{|x - x_0| \le a, \quad |y - y_0| \le b\}$$

with centre at the initial point (x_0, y_0); now let us attempt to determine the integral curve $y(x)$ which starts at the given initial point (x_0, y_0) and goes in the direction of increasing $x > x_0$.

Assume that in D the function $f(x, y)$ is continuous and satisfies the Lipschitz condition with respect to the variable y

$$|f(x, y_1) - f(x, y_2)| \le N|y_1 - y_2|, \tag{2.39}$$

where N is a constant depending neither on x nor on y. Note that the continuity of the function $f(x, y)$ in the closed domain D implies its boundedness in the given domain

$$|f(x, y)| \le M, \quad (x, y) \in D. \tag{2.40}$$

The required integral curve, if it exists, will intersect either the vertical line $x = x_0 + a$ or the horizontal one $y = y_0 + b$ or $y = y_0 - b$ on the boundary of the domain D. In the latter case the abscissa of the point of intersection is less than $x_0 + a$ and the required integral curve is no longer defined on the entire segment $x_0 \le x \le x_0 + a$. However, it follows from simple geometric considerations (see

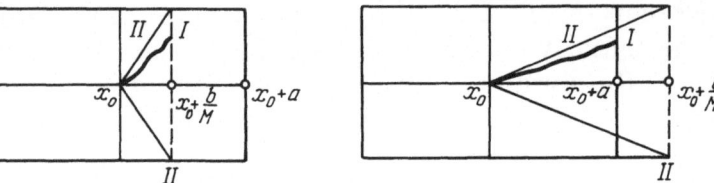

Fig. 3. $b/M \leq a$ Fig. 4. $b/M \geq a$

I. Integral curve passing through the point (x_0, y_0); II. Lines with slope $\pm M$

Figs. 3,4) that it will not intersect the horizontal boundaries until the point $x_0 + \dfrac{b}{M}$. Therefore, from now on, instead of the domain D, we shall consider the rectangle

$$\Delta = \{|x - x_0| \leq H, |y - y_0| \leq b\},$$

where $H = min\ \{a, b/M\}$.

Now let us turn to the construction of the Euler polygonal line. Divide the segment $[x_0, X]$, $X = x_0 + H$ into n parts by means of the partition points $x = x_0, x_1, \ldots, x_n = X$. Denote $x_i - x_{i+1} = h_i$ and $h = \max \{h_i\}$. At the first step, let us "freeze" $f(x, y)$ at the point (x_0, y_0), i. e. replace the right-hand side of (2.38) by the value $f(x_0, y_0)$. Then we shall obtain an equation with constant right-hand side

$$\frac{d\bar{y}}{dx} = f(x_0, y_0), \tag{2.41}$$

whose integral curve will be a segment of the straight line

$$\bar{y}(x) = y_0 + f(x_0, y_0)\ (x - x_0), \quad x \in [x_0, x_1]. \tag{2.42}$$

At the point x_1 this solution assumes the value $y_1 = y_0 + f(x_0, y_0)(x_1 - x_0)$. At the second step, let us take (x_1, y_1) for the new initial point and again, freezing $f(x, y)$ at this point, construct the next rectilinear segment, etc.

The polygonal line thus obtained is called the Euler polygonal line. It is easy to see that on the segment $[x_0, X]$ the Euler polygonal line will not leave the rectangle Δ. Indeed, it follows from (2.42) that the function $\bar{y}(x)$ satisfies the following inequality when $x \in [x_0, x_1]$

$$|\bar{y}(x) - y_0| \leq M(x - x_0) \leq MH \leq b,$$

i. e. the first segment of the polygonal line does not leave Δ. Therefore $|f(x_1, y_1)| \leq M$. Then, at the following step, for $x \in [x_1, x_2]$, we have

$$|\bar{y}(x) - y_0| \leq |\bar{y}(x) - y_1| + |y_1 - y_0| = |f(x_1, y_1)|\ (x - x_1) + |y_1 - y_0|$$

$$\leq M(x - x_1) + M(x_1 - x_0) = M(x - x_0) \leq MH \leq b$$

i. e. the second segment of the Euler polygonal line will not leave Δ. Continuing in a similar way, we see that the entire Euler polygonal line does not leave Δ. For a sufficiently small step in our partition, it is natural to consider the Euler polygonal line as an approximate integral curve. Thus we have obtained an algorithm for constructing an approximate solution of the initial value problem. In order to establish the validity of the algorithm described above, it is sufficient to prove that the sequence of Euler polygonal lines $\{\bar{y}_{(n)}(x)\}$ converges as $h \to 0$ and the limiting function $y(x)$ is the solution of the initial value problem (2.38). This is, at the same time, a proof of the existence theorem of the initial value problem's solution.

Definition. Any function $\tilde{y}(x)$, continuous on the segment $[x_0, X]$ and possessing a piecewise continuous derivative $\dfrac{d\tilde{y}}{dx}$ whose graph is entirely contained in Δ, is said to be an *ε-approximation with respect to discrepancy* of the solution of the initial value problem (2.38) if $|\tilde{y}(x_0) - y_0| \le \varepsilon$ and, when the function $\tilde{y}(x)$ is substituted into the equation (2.38), the latter assumes the form

$$\frac{d\tilde{y}}{dx} = f(x, \tilde{y}) + \psi(x) \tag{2.43}$$

where the discrepancy $\psi(x)$ satisfies the inequality

$$\sup_{x \in [x_0, X]} |\psi(x)| \le \varepsilon. \tag{2.44}$$

Obviously, the exact solution of the initial value problem, if it exists, may be viewed as an ε-approximation with respect to discrepancy of the solution for $\varepsilon = 0$.

Suppose that for any $\varepsilon > 0$ there exists an ε-approximation with respect to discrepancy of the solution of the initial value problem (2.38). Then we have the following lemma.

Lemma 2.2. *For any sufficiently small $\varepsilon > 0$, there is an $\varepsilon_1 > 0$ such that all the ε_1-approximations with respect to discrepancy of the solution of problem (2.38) differ from each other on the segment $[x_0, X]$ by no more than ε.*

Proof. Consider two arbitrary ε_1-approximations $\tilde{y}_{(1)}(x)$ and $\tilde{y}_{(2)}(x)$ with respect to discrepancy of the solution of problem (2.38). Obviously

$$\frac{d}{dx}\tilde{y}_{(1)} = f(x, \tilde{y}_{(1)}(x)) + \psi_1(x) \tag{2.45}$$

$$\frac{d}{dx}\tilde{y}_{(2)} = f(x, \tilde{y}_{(2)}(x)) + \psi_2(x), \tag{2.46}$$

where

$$|\tilde{y}_{(1)}(x_0) - \tilde{y}_{(2)}(x_0)| \le 2\varepsilon_1 \tag{2.47}$$

$$\sup_{x \in [x_0,\, X]} |\psi_1(x) - \psi_2(x)| \le 2\varepsilon_1. \tag{2.48}$$

Denote

$$\tilde{y}_{(2)}(x) - \tilde{y}_{(1)}(x) = z(x) \tag{2.49}$$

$$\psi_2(x) - \psi_1(x) = \varphi(x). \tag{2.50}$$

Subtracting (2.45) from (2.46), we obtain

$$\frac{dz}{dx} = f(x, \tilde{y}_{(2)}(x)) - f(x, \tilde{y}_{(1)}(x)) + \varphi(x) \tag{2.51}$$

Since $\tilde{y}_{(1)}(x)$ and $\tilde{y}_{(2)}(x)$ do not leave the rectangle Δ where the function $f(x, y)$ satisfies the Lipschitz condition (2.39), we obtain (from (2.51)) the differential inequality

$$\left|\frac{dz}{dx}\right| \le N|z| + 2\varepsilon_1 \tag{2.52}$$

In view of Lemma 2.1, we have

$$|z(x)| \le s(x) \tag{2.53}$$

where $s(x)$ is the solution of the initial value problem

$$\frac{ds}{dx} = Ns + 2\varepsilon_1, \quad s(x_0) = |z(x_0)|. \tag{2.54}$$

Using the explicit form (2.35) of the function $s(x)$, we get

$$|z(x)| \le |z(x_0)| e^{N(x-x_0)} + \frac{2\varepsilon_1}{N} (e^{N(x-x_0)} - 1).$$

Since the estimate (2.47) implies $z(x_0) \le 2\varepsilon_1$, this inequality may be rewritten in the form

$$|z(x)| = |\tilde{y}_{(1)}(x) - \tilde{y}_{(2)}(x)| \le 2\varepsilon_1 \left[e^{N(x-x_0)} + \frac{1}{N} (e^{N(x-x_0)} - 1) \right]. \tag{2.55}$$

Hence

$$\sup_{x \in [x_0,\, X]} |\tilde{y}_{(2)}(x) - \tilde{y}_{(1)}(x)| \le 2\varepsilon_1 \left[e^{NH} + \frac{1}{N} (e^{NH} - 1) \right] = 2\varepsilon_1 \Omega \tag{2.56}$$

where $\Omega > 0$ is a constant not depending on ε_1. Choosing $\varepsilon_1 < \dfrac{\varepsilon}{2\Omega}$, we obtain the statement of the lemma.

Suppose $\{\tilde{y}_{(n)}(x)\}$ is a sequence of ε_n-approximations with respect to discrepancy, i. e.

$$\sup_{x \in [x_0, \, X]} |\psi_n(x)| \le \varepsilon_n, \quad |\tilde{y}_{(n)}(x) - y_0| \le \varepsilon_n \tag{2.57}$$

Definition. If $\lim\limits_{n \to \infty} \varepsilon_n = 0$, then the sequence $\{\tilde{y}_n(x)\}$ will be called convergent with respect to discrepancy.

Since the constant Ω in the estimate (2.56) does not depend on x, it follows from Lemma (2.2) that a sequence converging with respect to discrepancy will converge uniformly on $[x_0, X]$.

We have the following main theorem.

Theorem 2.1. *Suppose that the function $f(x, y)$ in the initial value problem $\dfrac{dy}{dx} = f(x, y)$, $y(x_0) = y_0$, is continuous and satisfies the Lipschitz condition with respect to the variable y in the domain D. Then the sequence $\tilde{y}_{(n)}(x)$, which converges with respect to discrepancy on the segment $[x_0, X]$, will converge uniformly on this segment to a function $y(x)$ which will be a solution of the initial value problem.*

Proof. In view of Lemma 2.2, the sequence $\{\tilde{y}_{(n)}(x)\}$ satisfies the Cauchy criterion for uniform convergence on the closed interval $[x_0, X]$. Thus there exists a function $y(x)$ to which the sequence $\{\tilde{y}_{(n)}(x)\}$ converges uniformly and this function will be continuous since the $\tilde{y}_{(n)}(x)$ are continuous.

Substituting the ε_n-approximate solution $\tilde{y}_{(n)}(x)$ into Eq. (2.38) and replacing the identity thus obtained by an equivalent integral relation (see Lemma 1.1), we obtain

$$\tilde{y}_{(n)}(x) = \tilde{y}_{(n)}(x_0) + \int_{x_0}^{x} [f(\xi, \tilde{y}_{(n)}(\xi)) + \psi_n(\xi)] d\xi \tag{2.58}$$

Since $|\psi_n| \le \varepsilon_n$ and $|\tilde{y}_{(n)}(x_0) - y_0| \le \varepsilon_n$, passing to the limit when $\varepsilon_n \to 0$ in (2.58), we get

$$y(x) = y_0 + \int_{x_0}^{x} f(\xi, y(\xi)) d\xi. \tag{2.59}$$

It follows from this last identity that the limiting function $y(x)$ is differentiable. Taking derivatives, we find

$$\frac{dy}{dx} = f(x, y). \tag{2.60}$$

Moreover, $y(x_0) = y_0$. Thus the limiting function of the sequence $\{\tilde{y}_{(n)}(x)\}$ is an exact solution of the initial value problem (2.38). The theorem is proved.

In order to prove the existence theorem for solutions of the initial value problem (2.38), it now remains to show that there exists a sequence of ε_n-approximations (with repect to discrepancy) of the solution of this problem and that the sequence converges with respect to discrepancy. Now we shall show that the Euler polygonal lines constructed above constitute such a sequence.

Theorem 2.2. *Under the assumptions of Theorem 2.1, the sequence of Euler polygonal lines converges uniformly on the closed interval $[x_0, X]$, as $h \to 0$, to the function $y(x)$ which is the solution of the initial value problem.*

Proof. Since the initial values of the Euler polygonal lines $\bar{y}_{(n)}(x)$ all coincide by construction with y_0, it is sufficient to verify that as $h \to 0$ the discrepancies $\psi_n(x)$ converge to zero uniformly on $[x_0, X]$.

Substituting the Euler polygonal line $\bar{y}_{(n)}(x)$ (for an arbitrary step $x_s \le x \le x_{s+1}$) into equation (2.38), we obtain the following expression for the discrepancy $\psi_n(x)$:

$$\psi_n(x) = \frac{d}{dx}\bar{y}_{(n)}(x) - f(x, \bar{y}_{(n)}(x))$$

$$= f(x_s, \bar{y}_{(n)}(x_s)) - f(x, \bar{y}_{(n)}(x)), \quad x \in [x_s, x_{s+1}]. \tag{2.61}$$

Since $|x - x_s| \le h$ and $\bar{y}_{(n)}(x) = \bar{y}_{(n)}(x_s) + (x - x_s)\, f(x_s, y_{(n)}(x_s))$, $x \in [x_s, x_{s+1}]$, we have the estimate

$$\left|\bar{y}_{(n)}(x) - \bar{y}_{(n)}(x_s)\right| \le Mh \tag{2.62}$$

In view of the uniform continuity of the function $f(x, y)$, this indeed implies that for any sufficiently small $\varepsilon > 0$ there is an $h_0(\varepsilon)$ such that $h < h_0(\varepsilon)$ implies

$$\sup_{x \in [x_0, X]} |\psi_n(x)| \le \varepsilon, \tag{2.63}$$

which proves (in view of theorem 2.1) the uniform convergence (on the closed interval $[x_0, X]$) as $h \to 0$ of the sequence of Euler polygonal lines to the function $y(x)$ which is the solution of the initial value problem. The theorem is proved.

The theorems proved above imply the following statements.

Theorem 2.3 (*existence theorem*). *If the function $f(x, y)$ in the domain D is continuous and satisfies the Lipschitz conditions with respect to the variable y, then the solution of the initial value problem exists on the closed interval $[x_0, X]$.*

Under these assumptions concerning the function $f(x, y)$, the solution of the initial problem (2.38) is unique. We have the following.

Theorem 2.4 (*uniqueness theorem*). *Under the assumptions of Theorem 2.3, the initial value problem (2.38) possesses a unique solution on $[x_0, X]$.*

This theorem may be viewed as a consequence of Lemma 2.2. If we assume that there are two exact solutions of the Cauchy problem, then their initial values coincide and their discrepancy equals zero. Hence, according to Lemma 2.2, these solutions coincide completely on the closed interval $[x_0, X]$.

Besides the notion of ε-approximation with respect to discrepancy introduced above, one often uses the notion of an approximate solution with respect to deviation. Let us give its definition.

Definition. A function $\tilde{y}(x)$, bounded on $[x_0, X]$, is said to be an ε-*approximation with respect to deviation* of a solution of the Cauchy problem (2.38), if the exact solution $y(x)$ of the Cauchy problem exists and

$$\sup_{x \in [x_0, X]} |\tilde{y}(x) - y(x)| \leq \varepsilon \qquad (2.64)$$

It follows from our previous arguments that the convergence of a sequence of approximate solutions with respect to discrepancy implies its convergence with respect to deviation.

Note that the converse statement is not always true: if the deviation of the approximate solutions from the exact solution tends to zero, the solutions themselves may possess discrepancies which are as large as we wish; moreover, solutions which are approximate with respect to deviation may not necessarily be differentiable and may even be discontinuous.

Remarks. 1. The Euler polygonal line method not only yields a proof of the existence of solutions of the initial value problem, but also gives an effective algorithm for constructing approximate solutions which approximate the exact solution as closely as we wish. This method may easily be worked out on a computer. Thus it is one of the most effective numerical methods for finding solutions of the initial value problem. When it is actually carried out, a number of questions arise: they concern estimates of the error in the approximations obtained, as well as certain specific computer-oriented aspects of numerical methods. These questions will be considered in more detail in Chapter 6.

2. We have proved the existence and the uniqueness of the solution $y(x)$ of the initial value problem (2.38) only on the closed interval $[x_0, X]$. Suppose the function $f(x, y)$ is continuous and satisfies the Lipschitz conditions (2.39) in the given domain G. Since for $x = X$ the integral curve has not come out of the domain G, choosing a point $x = X, y = y(X)$ for the initial point we can repeat the arguments carried out previously and continue the solution to a new segment $[X, X_1]$. It can be shown that, by continuing this process, it is possible to construct an integral curve which approaches the boundary of the domain G as closely as we wish.

We have constructed an algorithm for finding the integral curve $y(x)$ in the direction of increasing values $x > x_0$. Obviously similar arguments may be carried out for constructing an integral curve in the direction of decreasing values $x < x_0$.

Here also the corresponding process of continuing the integral curve may be carried out until the integral curve reaches the boundary of the domain G.

3. It is possible to prove the existence of the solution of the initial value problem under the single requirement of the continuity of the function $f(x,y)$ (Peano's theorem). However the continuity of the function $f(x,y)$ is insufficient for the uniqueness of the solution of the initial value problem. Thus, for example, the problem

$$\frac{dy}{dx} = \sqrt{y}, \quad y(0) = 0, \tag{2.65}$$

besides the trivial solution $y \equiv 0$, also possesses the solution

$$y = \frac{x^2}{4}, \tag{2.66}$$

which satisfies the zero initial condition. (As can be easily seen, the right-hand side of equation (2.65) in a neighbourhood of the point $(0,0)$ possesses unbounded derivatives and does not satisfy the Lipschitz condition.)

4. If there is a unique integral curve passing through the point (x_0, y_0) and this curve is a solution of the problem (2.38) for the given differential equation, then the point (x_0, y_0) is said to be an ordinary point of the given equation. A point (x_0, y_0) of the domain G which is not an ordinary point is said to be a singular point of the given differential equation. For a singular point there are either no integral curves passing through it, or at least two integral curves which pass through it (an infinite set of integral curves may pass through a singular point).

If the assumptions of the existence and the uniqueness theorem hold in the neighbourhood of the point (x_0, y_0), then this point will be ordinary. It should be kept in mind that if the assumptions of the existence and uniqueness theorem stated above do not hold, the point is not necessarily singular; their violation is only a necessary condition, not a sufficient one for the given point to be singular. Therefore, in order to find out if the given point in singular or not, it is necessary to carry out an additional study.

5. At the beginning of §1 in Chapter 2, it was indicated that in many cases one should consider, together with the equation (2.1), the equation (2.2). If the assumptions of Theorem 2.3 for (2.1) break down at the point (x_0, y_0) because $f(x,y)$ becomes infinite, then $1/f(x,y)$ vanishes at this point and for equation (2.2) the requirements of the existence and the uniqueness theorem are met. Thus, in this case, the point (x_0, y_0) is an ordinary one, but the integral curve which passes through it has a vertical tangent at this point.

6. If the function $f(x,y)$ is continuous in the domain G and possesses discontinuities of the first kind on the line $x = x_k = \mathrm{const}$ $(k = 1, 2, \ldots, N)$ then, even under the assumption that $f(x,y)$ satisfies the requirements of Theorem 2.3 in the places where it is continuous, there exists no ordinary (or, as it is called, "classical") solution of the initial value problem (2.38) in the domain G.

However, as can easily be seen, it is possible to apply the algorithm for the succesive construction of the Euler polygonal lines $\{\bar{y}_{(n)}(x)\}$ in this case. The discontinuity points x_k will always be included in the set of partition points of the closed interval $[x_0, X]$ and, in the role of the fixed value of the function $f(x, y)$ at the step which begins at a discontinuity point, we shall always choose a definite limiting value (say the right-hand side one) of the function $f(x, y)$. Then the limiting function $y(x)$ of the sequence $\{\bar{y}_{(n)}(x)\}$ when $h\to 0$ will obviously turn out to be continuous, with a piecewise continuous derivative possessing discontinuities of the first kind on the vertical lines $x = x_k$. If we substitute the function $y(x)$ into the given differential equation, the discrepancy on the intervals of continuity of $f(x, y)$ is equal to zero. This limiting function $y(x)$ is a generalized solution of the initial value problem (2.38) on the closed interval $[x_0, X]$; the notion of a generalized solution was discussed in §1 of Chapter 1.

§3. Equations not Resolved with Respect to the Derivative

1. Theorem on the existence and uniqueness of the solution. Now let us consider the differential equation of first order of general form

$$F(x, y, y') = 0 \tag{2.67}$$

and determine sufficient conditions for the existence of solutions of this equation. The function F in its domain of definition gives a relation between the unknown function y, its derivative $y' = \dfrac{dy}{dx}$ and the independent variable x. If this relation could be resolved with respect to the derivative y', then we would obtain one or several differential equations of the first order resolved with respect to the derivative

$$y' = f_k(x, y) \quad (k = 1, 2, \ldots). \tag{2.68}$$

Suppose the function $f_k(x, y)$ in the neighbourhood of the point (x_0, y_0) in the (x, y)-plane satisfies the assumptions of the existence and uniqueness theorem for the solution of the Cauchy initial value problem for first order equations resolved with respect to the derivative. Then there is one and only one integral curve $y_k(x)$ for each of these equations $(k = 1, 2, \ldots)$ passing through the point (x_0, y_0). All these integral curves are solutions of the given differential equation (2.67) (when we substitute the functions $y_k(x)$ in the equation (2.67), we obtain an identity). The direction of the tangent vector to the integral curve $y_k(x)$ of the equation (2.68) at the point (x_0, y_0) is determined by the value of the function $f_k(x_0, y_0)$. If these values differ, then several integral curves of the equation (2.67) pass through the point (x_0, y_0) (namely, as many as there are equations of the form (2.68) obtained by resolving the equation (2.67) with respect to the derivative), but the directions of the tangent vectors to these curves at the point (x_0, y_0) differ. Therefore, in order to specify a definite solution of equation (2.67), it is necessary not only to choose initial data, i. e. the value of the solution $y(x)$ at the point x_0,

$$y(x_0) = y_0, \tag{2.69}$$

but also to specify the value of the derivative of the solution at this point: $y'(x_0) = y_0'$. Obviously, this value may not be chosen arbitrarily: y_0' must be a root of the equation

$$F(x_0, y_0, y') = 0. \tag{2.70}$$

Thus the existence of the solution of equation (2.67) is related to the possibility of resolving it with respect to y' and to the existence of the solution of equations (2.68). Thus a sufficient condition for the existence of a solution of equation (2.67) may be found by using the conditions for the existence of an implicit function and its continuity together with its derivative, which are well known from the calculus course. We have the following theorem.

Theorem 2.5 (*existence and uniqueness*). *Suppose that in some closed three-dimensional rectangle D_3 with centre at the point (x_0, y_0, y_0'), where y_0' is the real root of the equation $F(x_0, y_0, y') = 0$, we have the following conditions:*
   a) $F(x, y, y')$ is continuous with respect to the set of its variables together with the partial derivatives $\dfrac{\partial F}{\partial y}$ and $\dfrac{\partial F}{\partial y'}$,
   b) $\dfrac{\partial F}{\partial y'}(x_0, y_0, y_0')] \neq 0$.
Then in the neighbourhood of the point $x = x_0$ there exists a solution $y = y(x)$ of equation (2.67) satisfying the conditions

$$y(x_0) = y_0, \quad y'(x_0) = y_0', \tag{2.71}$$

and this solution is unique.

   Proof. By the assumptions a) and b) of the theorem, we have the conditions for existence and uniqueness of the implicit function

$$y' = f(x, y), \tag{2.72}$$

in a neighbourhood of the point (x_0, y_0, y_0') satisfying the condition

$$y_0' = f(x_0, y_0), \tag{2.73}$$

and there exists a closed rectangle D_2 with centre at the point (x_0, y_0) in which the function $f(x, y)$ is continuous together with its derivative $\dfrac{\partial f}{\partial y}$, which may be computed according to the chain rule

$$\frac{\partial f}{\partial y} = -\frac{\dfrac{\partial F}{\partial y}(x,y,f(x,y))}{\dfrac{\partial F}{\partial y'}(x,y,f(x,y))}. \tag{2.74}$$

But this means that the initial value problem $y(x_0)=y_0$ for the equation (2.72) possesses a unique solution on the closed interval

$$|x-x_0| \le H, \tag{2.75}$$

so that all the assumptions of the existence and the uniqueness theorems 2.1 and 2.2 are met. Theorem 2.5 is proved.

If the integral curves of equations (2.68) which intersect at the point (x_0, y_0) possess a common tangent (at this point) whose direction is determined by the value y_0', then the uniqueness conditions for the resolubility at this point of equation (2.67) with respect to y' will obviously break down at this point.

Example 2.3. Consider the equation

$$(y')^2 - (2x+y)y' + 2xy = 0. \tag{2.76}$$

Solving it with respect to the derivative y', we obtain two first order equations resolved with respect to the derivative

$$y' = y, \tag{2.77}$$

$$y' = 2x, \tag{2.78}$$

whose right-hand sides satisfy the existence and uniqueness assumptions (theorem 2.1 and 2.2) for the solution of the initial value problem at any point of the (x, y)-plane. The general solutions of equations (2.77) and (2.78) are of the form

$$y = C_1 e^x \tag{2.79}$$

and

$$y = x^2 + C_2, \tag{2.80}$$

where the constants C_1 and C_2 can be determined from the initial conditions. It is clear that an integral curve of the family (2.79), as well as an integral curve from the family (2.80), passes through any point of the (x, y)-plane and the curves of these families possess a common tangent $y'(x_0)=2x_0$ at the points of the line $y=2x$ (Fig. 5). (At the point $(0, 0)$ the integral curve $y=x^2$ is tangent to the line $y \equiv 0$ which is the particular solution of Eq. (2.79) obtained from formula (2.79)

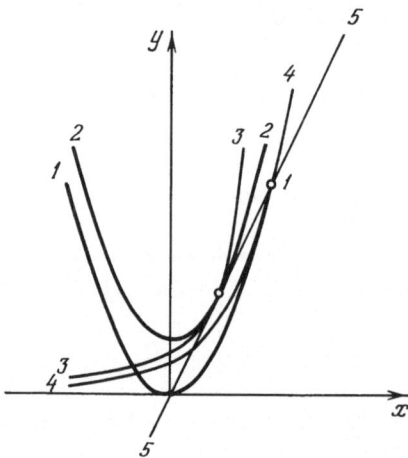

Fig. 5. 1. the curve $y=x^2$; 2. the curve $y=x^2+1$; 3. the curve $y=(2/e)e^x$; 4. $y=(2/e)^2e^x$;
5. the line $y=2x$

by putting $C_1=0$. This same integral curve $y=x^2$ intersects the integral curve
$y=(2/e)^2e^x$ of the family (2.79) at the point $x=2$, $y=4$ and both curves possess
the common tangent $y'=4$ at this point). Thus the line $y=2x$ is the geometric
locus of points where the assumptions of the uniqueness theorem for the solution
of equation (2.76) break down. (At these points condition b) fails to hold since
$\dfrac{\partial F}{\partial y'}\bigg|_{y=2x}=0$).

**2. The integration of an equation unresolved with respect to the derivative by
introducing a parameter.** Theorem 2.5, proved in the previous subsection,
guarantees, under certain conditions, the possibility of reducing the given
equation (2.67) to Eq. (2.68) and the solubility of the latter. However, the actual
realization of this possibility and the successful integration of equation
(2.72) thus obtained often meets with considerable difficulties. Therefore, in
many cases, it is more convenient to have other methods for integrating equation
(2.67). Let us begin with the case when Eq. (2.67) can easily be resolved with
respect to the unknown function itself

$$y(x)=f(x,y'). \tag{2.81}$$

For the sequel it is convenient to introduce the notation $y'=p$ and rewrite
(2.81) in the form

$$y(x)=f(x,p(x)). \tag{2.82}$$

Assuming the existence of the solution $y(x)$ of the given equation (2.67), we can differentiate relation (2.82) with respect to the independent variable x. We then obtain

$$\frac{dy}{dx} = p(x) = \frac{\partial f}{\partial x} + \frac{\partial f}{\partial p}\frac{dp}{dx}. \qquad (2.83)$$

The relation written above is a first order differential equation resolved with respect to $\dfrac{dp}{dx}$. The general solution of (2.83) may be written in the form of a one-parameter family

$$p(x) = \varphi(x, C). \qquad (2.84)$$

Hence, by using (2.82), we can obtain the family of solutions of the given equation (2.67) in the form

$$y = f(x, \varphi(x, C)) \qquad (2.85)$$

and, in order to solve the initial value problem, it remains for us to determine the value of the constant C by using the initial conditions.

Example 2.4. Consider the equation

$$(y')^2 - xy' + y = 0. \qquad (2.86)$$

It is obvious that this equation can easily be rewritten in the form (2.82):

$$y = xp - p^2, \qquad (2.87)$$

so that $p = p + (x - 2p)\dfrac{dp}{dx}$, i.e.

$$(x - 2p)\frac{dp}{dx} = 0. \qquad (2.88)$$

Equation (2.88) has the family of solutions

$$p(x) - C \qquad (2.89)$$

and, moreover, the solution

$$p(x) = \frac{x}{2}. \qquad (2.90)$$

Thus, taking into consideration (2.87), we obtain the solution of the given equation (2.86) in the form

$$y(x) = Cx - C^2 \qquad (2.91)$$

and

$$y(x) = \frac{x^2}{4}. \qquad (2.92)$$

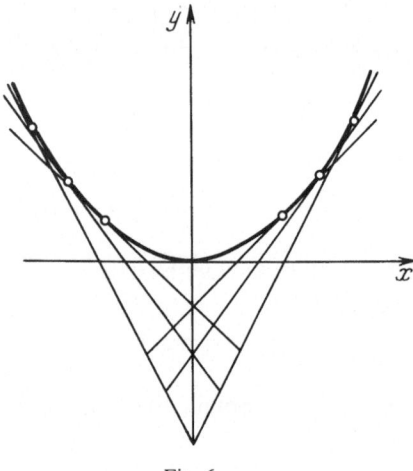

Fig. 6

It is easy to check that, for any point (x_0, y_0) belonging to the domain where the solution of equation (2.86) exists, two distinct integral curves (2.91), corresponding to two values of the constant C, pass through it (Fig. 6):

$$C = \frac{x_0}{2} \pm \sqrt{\frac{x_0^2}{4} - y_0}. \tag{2.93}$$

In order to specify a unique solution (of the initial value problem) passing through the point (x_0, y_0), we must also choose the value $y'(x_0) = y_0'$ determining the direction of the tangent to the integral curve at this point. It is also easy to see that the solution (2.92) of equation (2.86) possesses the following property: at each of its points, the curve $y = x^2/4$ is tangent to one of the curves (2.91). This means that the curve $y = x^2/4$ is the locus of all the points through which two solutions of equation (2.86) pass, the solutions having a common tangent at this point. At the points of the curve $y = x^2/4$ condition b) of Theorem 2.5 breaks down:

$$\frac{\partial F}{\partial y'}\bigg|_{y = x^2/4} = 2y' - x\big|_{y = x^2/4} = 0. \tag{2.94}$$

Thus the solution $y = x^2/4$ turns out to be, in a certain sense, a singular solution of equation (2.86).

   Before we go on to the general case, note that since neither of the variables x or y was a preferred one in the given equation (2.67), the considerations described above remain valid in the case when the original equation can be resolved with respect to the independent variable x. For example, this will be the case when we consider the so-called Lagrange equation

$$x\varphi(y') + y\psi(y') = \chi(y'), \tag{2.95}$$

which is linear with respect to the variables x and y. A particular case of the Lagrange equation was the one considered in example (2.4).

Now let us turn to our exposition of the general method of integrating first order equations (2.67) not resolved with respect to the derivative by introducing a parameter. Denote $y'=p$ and rewrite equation (2.67) in the form

$$F(x, y, p) = 0. \tag{2.96}$$

Equation (2.96) determines a certain surface in three-dimensional space (x, y, p). As is known, the introduction of two parameters u, v enables one to represent the given surface in parametric form

$$x = X(u, v), \quad y = Y(u, v), \quad p = P(u, v). \tag{2.97}$$

In our case, the functions X, Y and P are related by equation (2.96) and the equality $dy = p \, dx$. From the last relation we obtain

$$\frac{\partial Y}{\partial u} \, du + \frac{\partial Y}{\partial v} \, dv = P(u, v) \left\{ \frac{\partial X}{\partial u} \, du + \frac{\partial X}{\partial v} \, dv \right\}. \tag{2.98}$$

It therefore follows that the parameters u and v cannot be independent. Suppose $v = v(u)$. Then it follows from (2.98) that the relationship between the parameters u and v is an ordinary differential equation with respect to the function v:

$$\frac{dv}{du} = \frac{P(u, v) \dfrac{\partial X}{\partial u} - \dfrac{\partial Y}{\partial u}}{\dfrac{\partial Y}{\partial v} - P(u, v) \dfrac{\partial X}{\partial v}}, \tag{2.99}$$

and this equation is resolved with respect to the derivative. The family of solutions of equation (2.99) may be written in the form

$$v = \varphi(u, C). \tag{2.100}$$

Then, by (2.97), we obtain a family of integral curves of equation (2.67) written in parametric form

$$x = X(u, \varphi(u, C)), \quad y = Y(u, \varphi(u, C)), \tag{2.101}$$

which solves the problem of integrating equation (2.67).

Obviously, in the case when the original equation can easily be resolved with respect to the variable y (the variable x)

$$y = f(x, p), \tag{2.102}$$

in the parametric representation (2.97) we must choose the remaining variables x, p (or y, p) for the parameters u, v:

$$x = x, \quad y = f(x, p), \quad p = p. \tag{2.103}$$

As may easily be checked, equation (2.99) thus obtained for determining the function $p(x)$ will coincide with equation (2.83).

3. Singular solutions of first order equations unresolved with respect to the derivative. In example 2.4 considered above, for the first order equation unresolved with respect to the derivative (2.86), we obtained the singular solution $y = x^2/4$ possessing the following property: at all its points the uniqueness of the solution for the Cauchy initial value problem breaks down. Now consider the conditions for the existence of a singular solution in the general case.

The set of points (x, y) in which the uniqueness of the solution of (2.67) is violated will be referred to as the singular set of this equation. It is clear that at least one of the assumptions of Theorem 2.5 breaks down at the points of the singular set. In most cases it is assumption b), i.e. we have $\dfrac{\partial F}{\partial y'} = 0$. Then, if assumption a) holds at the points of the singular set while simultaneously assumption b) is violated, we have the relations

$$F(x, y, p) = 0, \quad \frac{\partial F}{\partial p}(x, y, p) = 0. \tag{2.104}$$

Eliminating p from the relation (2.104), we obtain an implicit equation for the so-called *p-discriminant curve*

$$\Phi(x, y) = 0. \tag{2.105}$$

Any integral curve at all points (x, y) of which we have $\dfrac{\partial F}{\partial y'} = 0$ will be called a *singular solution*. If one of the branches of the p-discriminant curve turns out to be an integral curve of equation (2.67), then it must be a singular solution. Note that an arbitrary p-discriminant curve does not necessarily have to be a singular solution of equation (2.67): it may happen that it is not an integral curve of this equation. Thus, as can be easily verified in the case of example 2.3, the p-discriminant curve $y = 2x$ of equation (2.76) is not an integral curve, and therefore not a singular solution of this equation. The singular solution $y = x^2/4$ found in example 2.4 for the equation (2.86) is its p-discriminant curve.

In cases where the set of solutions of equation (2.67) may be written in the form of a one-parameter family

$$\Phi(x, y, C) = 0, \tag{2.106}$$

in which the value of the constant C determines various integral curves, and the family of functions (2.106) possesses the envelope $y=y(x)$ then, obviously, this envelope is also an integral curve of equation (2.67); an integral curve of the family (2.106) passes through each point of the envelope and this curve possesses a common tangent with the envelope at this point. Thus the envelope of the family of integral curves (2.106) is a singular solution of equation (2.67). As is known, an envelope of the one-parameter family (2.106) may be found by excluding parameter C from the relations

$$\Phi(x,y,C)=0, \quad \frac{\partial\Phi}{\partial C}(x,y,C)=0. \tag{2.107}$$

The curve $\Psi(x,y)=0$ obtained in this way is known as a *C-discriminant curve*. Thus the singular solution $y=x^2/4$ of equation (2.86) found in example 2.4 is (as can easily be checked) a C-discriminant curve of the family of solutions (2.91).

In conclusion, note that the system of relations (2.107) determines not only an envelope of the family (2.106), but also the set of multiple points of this family, in which the partial derivatives $\frac{\partial\Phi}{\partial x}$ and $\frac{\partial\Phi}{\partial y}$ either do not exist or simultaneously vanish. Therefore a condition for the existence of an envelope of the family (2.106) and thus for the existence of a singular solution of equation (2.67), is the existence of bounded partial derivatives $\frac{\partial\Phi}{\partial x}$ and $\frac{\partial\Phi}{\partial y}$ satisfying the condition

$$\left(\frac{\partial\Phi}{\partial x}\right)^2+\left(\frac{\partial\Phi}{\partial y}\right)^2\neq 0. \tag{2.108}$$

§4. Existence and Uniqueness Theorems for the Solution of Normal Systems

The main idea underlying the construction of Euler polygonal lines may be used to obtain a constructive proof of the existence of the solution not only in the case of one equation resolved with respect to the derivative, but also in the case of a normal system. These questions constitute the main contents of this section.

Thus let us consider the initial value problem for the normal system of first order equations

$$\frac{dy_i}{dt}=f_i(t,y_1,\ldots,y_m),$$

$$y_i(t_0)=y_i^0, \quad (i=1,2,\ldots,m). \tag{2.109}$$

Suppose the functions $f_i(t, y_1, \ldots, y_m)$ are defined in the domain D which is a $(m+1)$-dimensional cuboid

$$D = \{|t - t_0| \le a, \ |y_i - y_i^0| \le b_i (i = 1, 2, \ldots, m)\}$$

with centre at the point $(t_0, y_1^0, \ldots, y_m^0)$. Assume that in the domain D the functions $f_i(t, y_1, \ldots, y_m)$ are continuous (and therefore bounded) and satisfy the Lipschitz conditions with respect to the variables $y_1, \ldots, y_m$ i.e.

$$|f_i(t, y_1, \ldots, y_m)| \le M,$$

$$|f_i(t, \bar{y}_1, \ldots, \bar{y}_m) - f_i(t, y_1, \ldots, y_m)| \le N \sum_{j=1}^{m} |\bar{y}_j - y_j|. \tag{2.110}$$

where the constants M and N do not depend on i.

Repeating the arguments carried out in the case of one equation, it is easy to establish that the required integral curve (if it exists) will not leave the domain D on the closed interval $[t_0, T]$ where the independent variable t varies; here $T = t_0 + H$, while the value of H is determined as

$$H = \min \left\{ a, \ \frac{\min_i b_i}{M} \right\}. \tag{2.111}$$

In order to construct the Euler polygonal line on the closed interval $[t_0, T]$, let us subdivide this interval into n parts by partition points $t_0, t_1, \ldots, t_n = T$ and, just as we did in §2, denote $t_i - t_{i-1} = h_i$, $h = \max_i h_i$. At the first step let us "freeze" the functions $f_i(t, y_1, \ldots, y_n)$ at the point $(t_0, y_1^0 \ldots y_m^0)$ and, integrating the equations with constant right-hand sides thus obtained, find the values of the functions $\bar{y}_i(t)$ on the interval $[t_0, t_1]$

$$\bar{y}_i(t) = y_i^0 + f_i(t_0, y_1^0, \ldots, y_m^0) \ (t - t_0). \tag{2.112}$$

The functions found in this way determine (in the $(m+1)$-dimensional space of the variables $t, y_1, \ldots, y_m$) on the interval $[t_0, t_1]$ a certain rectilinear segment which will be the first segment of the Euler polygonal line.

Taking the values of the function $\bar{y}_i(t)$ at the point $t = t_1$ for the new initial conditions and, repeating the algorithm described above, we shall obtain an Euler polygonal line on the closed interval $[t_0, T]$. By means of estimates which are similar to those carried out for one equation in §2, we can prove that the Euler polygonal line constructed according to this algorithm on the closed interval $[t_0, T]$ will not leave the domain D.

Let us introduce the notion of ε-approximation with respect to discrepancy of the solution of the initial value problem (2.109), similar to the corresponding notion introduced in the case of one equation.

Definition. A vector-valued function $\tilde{y}(t) = (\tilde{y}_1(t), \ldots, \tilde{y}_m(t))$ with piecewise continuous derivatives $\dfrac{d\tilde{y}_i}{dt}$, which is continuous on the segment $[t_0, T]$ and whose graph is entirely contained in the domain D, is said to be an *ε-approximation with respect to discrepancy* of the solution of the initial value problem (2.109) if $\max\limits_i |\tilde{y}_i(t_0) - y_i^0| \leq \varepsilon$ and, when the functions $\tilde{y}_i(t)$ are substituted into equation (2.109):

$$\frac{d\tilde{y}_i}{dt} = f_i(t, \tilde{y}_1, \ldots, \tilde{y}_m) + \psi_i(t), \tag{2.113}$$

the discrepancies $\psi_i(t)$ satisfy the inequality

$$\max_i \sup_{t \in [t_0, T]} |\psi_i(t)| \leq \varepsilon. \tag{2.114}$$

In order to prove the existence theorem let us show, just as we did in the case of one equation, that the sequence of Euler polygonal lines $\{\tilde{y}_{(n)}(t)\}$, when $h \to 0$, constitutes the uniformly convergent sequence (on the closed interval $[t_0, T]$) of ε_n-approximations with respect to discrepancy of the solution of the initial value problem, while the limiting vector function of this sequence $\bar{y}(t)$ satisfies all the conditions of the original problem (2.109). This may be done by means of arguments similar to those used in §2.

Lemma 2.3. *For any $\varepsilon > 0$ there is an $\varepsilon_1 > 0$ such that all ε_1-approximations with respect to discrepancy of the solution of the initial value problem (2.109) differ from each other by no more than ε on the closed interval $[t_0, T]$.*

Proof. Choose two ε_1-approximations with respect to discrepancy of the solution of problem (2.109) $\tilde{y}_{(1)}(t)$ and $\tilde{y}_{(2)}(t)$. This means

$$\frac{d}{dt} \tilde{y}_{(1)i} = f_i(t, \tilde{y}_{(1)1}, \ldots, \tilde{y}_{(1)m}) + \psi_{(1)i}(t) \tag{2.115}$$

$$\frac{d}{dt} \tilde{y}_{(2)i} = f_i(t, \tilde{y}_{(2)1}, \ldots, \tilde{y}_{(2)m}) + \psi_{(2)i}(t), \tag{2.116}$$

where

$$\max_i |\tilde{y}_{(1)i}(t_0) - \tilde{y}_{(2)i}(t_0)| \leq 2\varepsilon_1 \tag{2.117}$$

$$\max_i \sup_{t \in [t_0, T]} |\psi_{(1)i}(t) - \psi_{(2)i}(t)| \leq 2\varepsilon_1. \tag{2.118}$$

Putting

$$\tilde{y}_{(2)i}(t) - \tilde{y}_{(1)i}(t) = z_i(t) \tag{2.119}$$

$$\psi_{(2)i}(t) - \psi_{1(i)}(t) = \varphi_i(t) \tag{2.120}$$

and subtracting (2.115) from (2.116), we get

$$\frac{dz_i}{dt} = f_i(t, \tilde{y}_{(2)1}, \ldots, \tilde{y}_{(2)m}) - f_i(t, \tilde{y}_{(1)1}, \ldots, \tilde{y}_{(1)m}) + \varphi_i(t). \tag{2.121}$$

Since $\tilde{y}_{(1)i}(t)$ and $\tilde{y}_{(2)i}(t)$ do not leave the domain D where the functions $f_i(t, y_1, \ldots, y_m)$ satisfy the Lipschitz condition (2.110), we have the estimate

$$\left| \frac{dz_i}{dt} \right| \leq N \sum_{j=1}^{m} |z_j(t)| + 2\varepsilon_1. \tag{2.122}$$

Introduce the function $\varrho(t) = \sqrt{\sum_{i=1}^{m} z_i^2(t)}$ and consider its derivative

$$\frac{d\varrho}{dt} = \frac{1}{\varrho} \sum_{i=1}^{m} z_i \frac{dz_i}{dt}.$$

Then, taking into consideration the fact that $|z_i| \leq \varrho$, the estimate (2.122) can be used to obtain

$$\left| \frac{d\varrho}{dt} \right| \leq \sum_{i=1}^{m} \left| \frac{dz_i}{dt} \right| \leq Nm^2 \varrho + 2m\varepsilon_1, \tag{2.123}$$

i.e.

$$\left| \frac{d\varrho}{dt} \right| \leq Nm^2 \varrho + 2m\varepsilon_1. \tag{2.124}$$

By lemma 2.1 we have the following estimate for ϱ

$$\varrho(t) \leq s(t),$$

where $s(t)$ is the solution of the initial value problem for the linear equation with constant coefficients

$$\frac{ds}{dt} = Nm^2 s + 2m\varepsilon_1, \quad s(t_0) = \varrho(t_0). \tag{2.125}$$

According to (2.117), the initial value of $s(t_0)$ satisfies the inequality

$$s(t_0) = \varrho(t_0) = \sqrt{\sum_{i=1}^{m} z_i^2(t_0)} \leq 2\sqrt{m}\,\varepsilon_1.$$

Therefore for $\varrho(t)$ we finally obtain

$$\varrho(t) \leq 2\sqrt{m}\,\varepsilon_1 e^{Nm^2(t-t_0)} + \frac{2\varepsilon_1}{Nm}\,(e^{Nm^2(t-t_0)} - 1)$$

$$\leq 2\varepsilon_1 \left[\sqrt{m}\,e^{Nm^2 T} + \frac{1}{Nm}\,(e^{Nm^2 T} - 1) \right] = 2\varepsilon_1 \Omega, \qquad (2.126)$$

where the constant Ω does not depend on ε_1. Obviously the estimate (2.126) is also valid for $|z_i(t)|\,(i=1,\ldots,m)$. Choosing $\varepsilon_1 = \varepsilon/2\Omega$, we obtain the statement of lemma 2.3.

Definition. A sequence of ε_n-approximate solutions $\tilde{y}_{(n)}(t)$ (with respect to discrepancy) is said to *converge with respect to discrepancy* if $\varepsilon_n \to 0$.

It follows from the lemma proved above that every sequence $\{\tilde{y}_{(n)}(t)\}$ which converges with respect to discrepancy is uniformly convergent on the closed interval $[t_0, T]$.

The following theorems can be proved similarly to the corresponding theorems for the case of one equation.

Theorem 2.6. *Suppose that in the initial value problem*

$$\frac{dy_i}{dt} = f_i(t, y_1, \ldots, y_m), \ \ y_i(t_0) = y_i^0 \,(i=1,\ldots,m)$$

the functions $f_i(t, y_1, \ldots, y_m)$ in the domain D are continuous and satisfy the Lipschitz conditions with respect to the variables $y_1, \ldots, y_m$. Then the sequence $\{\tilde{y}_{(n)}(t)\}$ converging with respect to discrepancy on the closed interval $[t_0, T]$ converges uniformly on this segment to a function $y(t)$ which is a solution of the initial value problem.

Theorem 2.7. *If the assumptions of Theorem 2.6 are satisfied, then the sequence of Euler polygonal lines converge uniformly as $h \to 0$ on the closed interval $[t_0, T]$ to the function $y(t)$ which is a solution of the initial value problem.*

This implies:

Theorem 2.8 *(existence). If the functions*

$$f_i(t, y_1, \ldots, y_m), \ i=1, \ldots, m,$$

are continuous in the domain D and satisfy the Lipschitz conditions with respect to the variables $y_1, \ldots, y_m$, then the solution of the initial value problem exists on the closed interval $[t_0, T]$.

Just as in the case of one equation, we have the following.

Theorem 2.9 *(uniqueness). Under the assumptions of Theorem (2.8), the initial value problem (2.109) possesses a unique solution on* $[t_0, T]$.

Thus the existence and uniqueness theorems for the solution of the initial value problem for a normal system have been proved entirely. Note that the remarks which we made in §2 concerning the existence and the uniqueness theorems of the solution of the initial value problem for one equation remain valid in the case of a normal system.

In Chapter 1, we showed that any n-th order equation (1.6) is equivalent to the normal system (1.8). Therefore it follows that if the right-hand side of equation (1.6) is a function $f\left(t, y, \dfrac{dy}{dt}, \ldots, \dfrac{d^{n-1}y}{dt^{n-1}}\right)$ satisfying the assumptions of theorem 2.8, then the solution of the initial value problem (1.6) exists and is unique.

It should be noted that the method for proving existence theorems (involving Euler polygonal lines) is the theoretical basis of effective algorithms for the numerical solution of initial value problems for sufficiently complicated systems of differential equations reduced to normal form. Later (see Chapter 6), we shall consider other algorithms for the numerical solution of differential equations which are more efficient from the practical point of view (they improve, for example, the rate of convergence of the approximations). Here we shall limit ourselves to an example of the numerical solution for a sufficiently complicated normal system, which was carried out practically only with the help of a large computer.

Consider the motion in interplanetary space of a spaceship acted on by the gravitation of the Earth, Moon, and Sun. The problem is to compute the

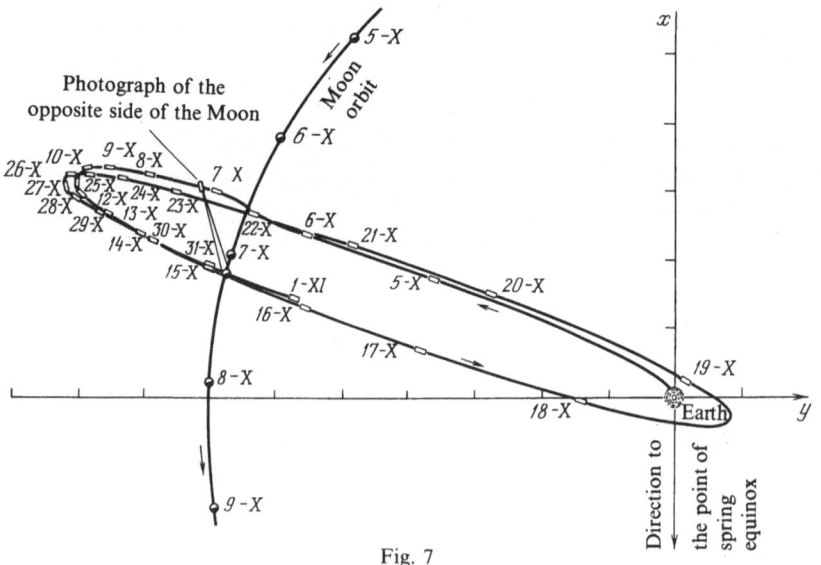

Fig. 7

trajectory of the spaceship to the Moon. The motion may be described by a system of equations for the motion of four bodies of type (1.20), where $F_i(i=1,2,3,4)$ is the resultant of the gravitational forces affecting the i-th body on the part of all the other bodies, where of course the action of the spaceship on the other three bodies is negligible. Thus we obtain a system of 12 equations of the second order or a normal system of 24 equations with right-hand sides which have a complicated analytical structure. The explicit formulas for the right-hand sides will not be written here. In order to apply the Euler algorithm and other numerical algorithms, it is sufficient to have the possibility of computing the right-hand sides for various positions of the moving bodies; thus the specific form of the analytic formulae for the right-hand sides is not too important for the integration method.

Figure 7 represented here shows one of the projections of the trajectory of motion of the automatic interplanetary station sent into space from the USSR on October 4, 1959 and used to photograph the opposite side of the Moon.

§ 5. Dependence of Solutions on Initial Values and Parameters

In the actual process of solving specific differential equations, the initial values are usually known only approximately, since they are determined experimentally or calculated, and this is always related to the appearance of errors. Moreover, the right-hand side of the equations may contain certain parameters which characterize the physical nature of the system under study (masses, charges, elastic characteristics etc.) whose values are also determined approximately. In this connection the following question arises: how will the solution of the initial value problem change for small changes of the initial values and parameters and does it depend continuously on them? This is the question which we shall consider in this section. Note that a similar question may be asked in the case of an unlimited interval $[t_0, \infty)$ if a solution exists on it. This question constitutes the contents of the so-called theory of stability, to which a special chapter is devoted (Chapter 5).

Let us consider the initial value problem for the normal system of differential equations

$$\frac{dy}{dt} = f(y, t, \mu),$$

$$y = (y_1, \ldots, y_m), \quad f = (f_1, \ldots, f_m) \tag{2.127}$$

with initial conditions

$$y(t_0) = y_0, \quad y_0 = (y_{10}, \ldots, y_{m0}). \tag{2.128}$$

Here $\mu = (\mu_1, \ldots, \mu_s)$ is a vector describing the parameters $\mu_1, \ldots, \mu_s$ involved in the right-hand side of the system.

We are interested in the character of the dependence of the solution on $y_{10}, \ldots, y_{m0}$ and $\mu_1, \ldots, \mu_s$. Note that our study of the dependence of the solutions on the initial conditions $y_{10}, \ldots, y_{m0}$ and on t_0 may be reduced to that of the dependence on the parameters in the right-hand side of the system. Indeed, carry out the change of variables

$$y_i = y_{i0} + z_i \quad (i = 1, \ldots, m), \quad t = t_0 + \tau \tag{2.129}$$

in (2.127) and write the obtained equations in the new unknown functions z_i:

$$\frac{dz_i}{d\tau} = \varphi_i(z, \tau, \mu, y_0, t_0) \quad (i = 1, \ldots, m),$$

$$\varphi_i(z, \tau, \mu, y_0, t_0) \equiv f_i(y_0 + z, t_0 + \tau, \mu). \tag{2.130}$$

For $t = t_0$ the new variable τ becomes equal to 0 and the initial values for z_i are now fixed

$$z_i(0) = y_i(t_0) - y_{i0} = 0. \tag{2.131}$$

The values y_{i0} and t_0 appear in the right-hand sides of (2.130) as parameters, together with the parameters $\mu_1, \ldots, \mu_s$. The problem thus reduces to studying the dependence of z_i on the parameters on the right-hand side. We also have the converse reduction: the study of the dependence on the parameters may be viewed as a certain particular case of the dependence of the solutions on the initial condition. Indeed, since the parameters $\mu_1, \ldots, \mu_s$ in (2.127) are fixed and assume, say, the values $\mu_{k0} (k = 1, \ldots, s)$, it follows that we may add an equation of the form $\dfrac{d\mu_k}{dt} = 0$ with initial condition $\mu_k(t_0) = \mu_{k0}$ to equation (2.127) with initial condition (2.128). We then obtain a new system

$$\frac{dy}{dt} = f(y, t, \mu), \quad \frac{d\mu}{dt} = 0, \quad y_i(t_0) = y_{i0}, \quad \mu_k(t_0) = \mu_{k0}. \tag{2.132}$$

Now the question of the dependence of y_i on μ_k reduces to the study of the dependence of the solution of problem (2.132) on the initial conditions $\mu_{10}, \ldots, \mu_{s0}$.

Therefore we shall study the dependence of the solutions on the parameters, and our conclusions on the dependence on initial conditions will be made on the basis of the equivalence which we have just established.

First consider the scalar equation

$$\frac{dy}{dt} = f(y, t, \mu) \tag{2.133}$$

with scalar parameter μ and fixed initial value

$$y(t_0) = y_0 \tag{2.134}$$

Suppose the right-hand side $f(y, t, \mu)$ defined in the cuboid

$$D = \{|t - t_0| \le a, \ |y - y_0| \le b, \ |\mu - \mu_0| \le C\}$$

is continuous in D as a function of the set of its variables and also satisfies the Lipschitz condition with respect to y in D:

$$|f(y_1, t, \mu) - f(y_2, t, \mu)| \le N|y_1 - y_2|, \tag{2.135}$$

where N is the same constant for all μ in the closed interval $|\mu - \mu_0| \le C$.

For every fixed μ, according to the existence theorem, an integral curve is defined on the interval

$$[t_0, t_0 + H] \left(H = \min \left\{ a, \frac{b}{M} \right\}, \ M = \sup_D |f(y, t, \mu)| \right)$$

and is the solution of the initial value problem (2.133), (2.134). If we change μ, and keep in mind the fact that the constants M and N do not depend on μ, it follows that a family of integral curves $y(t, \mu)$ will be defined on $[t_0, t_0 + H]$.

Let us study the dependence of $y(t, \mu)$ on μ. First we prove the following theorem.

Theorem 2.10. *Suppose $f(y, t, \mu)$ is defined and continuous in D and satisfies the Lipschitz condition (2.135) with respect to the variable y. Then the solution $y(t, \mu)$ of problem (2.133), (2.134) defined on the closed interval $[t_0, t_0 + H]$ is continuous with respect to μ for any μ in the closed interval $|\mu - \mu_0| \le C$.*

The theorem will be proved if we establish that $\forall \varepsilon > 0 \, \exists \delta(\varepsilon)$ such that for $|\Delta\mu| < \delta$ the following inequality

$$|y(t, \mu + \Delta\mu) - y(t, \mu)| < \varepsilon \tag{2.136}$$

holds for all μ and $\mu + \Delta\mu$ belonging to the closed interval $|\mu - \mu_0| \le C$. Using

Lemma 2.1 on differential inequalities, we see that

$$\frac{d}{dt} y(t, \mu + \Delta\mu) = f(y(t, \mu + \Delta\mu), t, \mu + \Delta\mu), \quad y(t_0, \mu + \Delta\mu) = y_0,$$

$$\frac{d}{dt} y(t, \mu) = f(y(t, \mu), t, \mu), \quad y(t_0, \mu) = y_0.$$

Subtracting one relation from the other, for the difference $\Delta y = y(t, \mu + \Delta\mu) - y(t, \mu)$ we obtain

$$\frac{d}{dt} \Delta y = [f(y(t, \mu + \Delta\mu), t, \mu + \Delta\mu) - f(y(t, \mu), t, \mu + \Delta\mu)]$$

$$+ [f(y(t, \mu), t, \mu + \Delta\mu) - f(y(t, \mu), t, \mu)], \quad \Delta y|_{t=t_0} = 0. \quad (2.137)$$

In view of the continuity of $f(y, t, \mu)$ with respect to the set of its variables and since $|y(t, \mu) - y_0| \le b$ for $|t - t_0| \le H$, $\forall \varepsilon_1 > 0$, $\exists \delta_1(\varepsilon_1)$ such that $|\Delta\mu| < \delta_1$ implies

$$|f(y(t, \mu), t, \mu + \Delta\mu) - f(y(t, \mu), t, \mu)| < \varepsilon_1$$

uniformly for all $t \in [t_0, t_0 + H]$. Using this fact and the Lipschitz conditions, we obtain

$$\left| \frac{d}{dt} \Delta y \right| < N |\Delta y| + \varepsilon_1. \quad (2.138)$$

Hence, according to Lemma 2.1 and formula (2.35), we get

$$|\Delta y| < \frac{\varepsilon_1}{N} (e^{N(t-t_0)} - 1) \le \frac{\varepsilon_1}{N} (e^{NH} - 1) < \varepsilon \quad (2.139)$$

whenever $\varepsilon_1 < \delta_2(\varepsilon)$, i.e. when $|\mu - \mu_0| < \delta_1(\delta_2(\varepsilon)) = \delta(\varepsilon)$. The theorem is proved.

Remarks. 1. It follows from the method of proof that inequality (2.136) holds uniformly with respect to t when $|t - t_0| \le H$, i.e. the function $y(t, \mu)$ is continuous in μ uniformly with respect to t, in other words, for a sufficiently small change of the parameter, the difference between two integral curves will be uniformly small on the entire closed interval under study $[t_0, t_0 + H]$.

2. Using the theorem proved above and a well-known theorem of calculus[1], it is easy to obtain the statement of the continuity of $y(t, \mu)$ as a function of the set

---

[1] Theorem. Suppose for $\alpha \in A$, $\beta \in B$ the function $y(\alpha, \beta)$ is continuous for all $\alpha \in A$. Then $y(\alpha, \beta)$ is a continuous function of the pair of arguments (α, β) whenever $\alpha \in A$, $\beta \in B$. Indeed, consider the difference $\Delta y = y(\alpha + \Delta\alpha, \beta + \Delta\beta) - y(\alpha, \beta)$; we have $|\Delta y| \le |y(\alpha + \Delta\alpha, \beta + \Delta\beta) - y(\alpha, \beta + \Delta\beta)| + |y(\alpha, \beta + \Delta\beta) - y(\alpha, \beta)|$; for all $\varepsilon > 0$ the first summand on the righthand side is less than $\varepsilon/2$ for $|\Delta\alpha| < \delta_1(\varepsilon)$ and all $\beta + \Delta\beta \in B$, while the second is less than $\varepsilon/2$ for $|\Delta\beta| < \delta_2(\varepsilon)$ and all $\alpha \in A$, so that $|\Delta y| < \varepsilon$ whenever $|\Delta\alpha| < \delta_1(\varepsilon)$ and $|\Delta\beta| < \delta_2(\varepsilon)$.

of its variables t, μ in the domain $|t-t_0|\le H$, $|\mu-\mu_0|\le C$. Indeed, the continuity of y with respect to μ is uniform with respect to t (as we have just proved), while the continuity of y in t is uniform with respect to μ (since $\left|\dfrac{dy}{dt}\right|=|f|\le M$).

3. If μ is now a vector and $f(y,t,\mu)$ is defined and continuous as a function of the set of its variables in

$$D=\{|t-t_0|\le a,\ |y-y_0|\le b,\ |\mu_k-\mu_k^0|\le C_k\}$$

and satisfies the Lipschitz condition with respect to y in D with a constant N not depending on μ, then the theorem just proved above may be applied, with μ in the role of one of the components of μ. The inequality (2.136) means that y is continuous with respect to this component, this continuity is uniform with respect to t and the other components of μ; then, according to remark 2, $y(t,\mu)$ is a continuous function of the set of its variables $t,\mu_1,\ldots,\mu_s$.

Now let us study the solution's dependence on the parameter y_0. Suppose this parameter varies in the closed interval $|y_0-y_0^0|\le\Delta$. Suppose $f(y,t)$ is defined and continuous in $D=\{|t-t_0|\le a,|y-y_0^0|\le b\}$. Then it is clear from geometrical considerations that the family of integral curves $y(t,y_0)$ exists on the closed interval $[t_0,t_0+\bar{H}]$ where $\bar{H}=\min\left\{a,\dfrac{b-\Delta}{M}\right\}$, $M=\sup\limits_{D}|f|$. Let us reduce this problem, using the method indicated above, to the previously studied problem of the solution's dependence on μ, by introducing $z=y-y_0$ and putting $y_0=\mu$. We then see, as a consequence of Theorem 2.10, that $y(t,y_0)$ is defined on $[t_0,t_0+\bar{H}]$ and continuous with respect to y_0 for any y_0 in the closed interval $|y_0-y_0^0|\le\Delta$.

Exactly in the same way, we can study the solution's dependence on the parameter t_0 which varies on the closed interval $|t_0-t_0^0|\le\delta$ and conclude that $y(t,t_0)$ is defined on the closed interval $[t_0^0,t_0^0+\bar{\bar{H}}]$, where $\bar{\bar{H}}=\min\left\{a-\delta,\dfrac{b}{M}-\delta\right\}$, and is continuous with respect to t_0 whenever $|t_0-t_0^0|\le\delta$.

If y is a vector, then similar results remain valid and may be proved by using the same Lemma 2.1 on differential inequalities and the considerations involved in the proof of the existence theorem for systems of differential equations.

Summarizing the above, we obtain the following results for the general case (2.127), (2.128). Suppose t_0,y_{i0} are parameters which vary in the domain $|t_0-t_0^0|\le\delta$, $|y_{i0}-y_{i0}^0|\le\Delta_i$ and the vector parameter μ appearing in the system varies in the domain $|\mu_k-\mu_k^0|\le C_k$. Suppose the right-hand sides of the system (2.127) are continuous in the set of their variables in the cuboid

$$D=\{|t-t_0^0|\le a,\ |y_i-y_{i0}^0|\le b_i,\ |\mu_k-\mu_k^0|\le C_k\} \tag{2.140}$$

and satisfy the Lipschitz condition with respect to $y_1,\ldots,y_m$ in D. Then the functions $y_i(t,t_0,y_{10},\ldots,y_{m0},\mu_1,\ldots,\mu_s)$ defined on the closed interval

$[t_0^0, t_0^0 + H]$, where

$$H = \min \left\{ a - \delta, \ \frac{\min (b_i - \Delta_i)}{\max M_i} - \delta \right\}, \quad M_i = \sup_D |f_i|,$$

are the solutions of the problem (2.127), (2.128), continuous in the set of their variables whenever $|t - t_0^0| \leq H$, $|t_0 - t_0^0| \leq \delta$, $|y_{i0} - y_{i0}^0| < \Delta_i$, $|\mu_k - \mu_k^0| \leq C_k$.

The property of continuity with respect to parameters is essential in viewing the initial value problem (2.127), (2.128) as the mathematical model for many scientific problems. Indeed, as we have already pointed out, in practice the initial data and parameters which appear on the right-hand sides of the equations are not known exactly: as a rule, they are only given approximately. However, in view of the theorem on the continuous dependence on parameters, a small change in initial data and right-hand sides of the equations of the system produce correspondingly small changes in the solution. This justifies the use of the solutions of problem (2.127), (2.128) to interpret the actual process whose mathematical model is the given system.

Let us now study the differentiability of solutions with respect to parameters and initial data. We shall consider this question for the scalar equation with scalar parameter μ. Besides the assumptions of Theorem 2.10, we shall require that $f(y, t, \mu)$ possess continuous partial derivatives $f_y(y, t, \mu)$ and $f_\mu(y, t, \mu)$ in the domain D.

If the solution $y(t, \mu)$ of problem (2.133), (2.134) possesses a derivative with respect to the parameter μ, then, substituting $y(t, \mu)$ into (2.133), (2.134) and taking the derivative of the identity thus obtained with respect to μ, we get

$$\frac{d}{dt} \frac{\partial y}{\partial \mu} = f_y(y(t, \mu), t, \mu) \frac{\partial y}{\partial \mu} + f_\mu(y(t, \mu), t, \mu), \tag{2.141}$$

$$\frac{\partial y}{\partial \mu}(t_0, \mu) = 0. \tag{2.142}$$

Let us check that under the supplementary assumptions imposed on $f(y, t, \mu)$ the derivative $\dfrac{\partial y}{\partial \mu}$ does indeed exist and satisfies equation (2.141) and initial condition (2.142).

To do this, consider the initial value problem

$$\frac{dz}{dt} = f_y(y(t, \mu), t, \mu)z + f_\mu(y(t, \mu), t, \mu), \tag{2.143}$$

$$z(t_0, \mu) = 0 \tag{2.144}$$

for the new unknown function $z(t, \mu)$. Equation (2.143) is linear with respect to z and its coefficients are continuous functions of t (in the interval $|t - t_0| \leq H$ where

$y(t,\mu)$ is continuous). Hence the solution $z(t,\mu)$ of the initial value problem (2.143), (2.144) may be written out explicitly according to formula (2.29) and is a continuous function of t when $|t-t_0|\leq H$.

Denote

$$\frac{\Delta y}{\Delta \mu}=\frac{y(t,\mu+\Delta\mu)-y(t,\mu)}{\Delta\mu}=w(t,\mu,\Delta\mu). \tag{2.145}$$

Let us check that $\lim_{\Delta\mu\to 0} w(t,\mu,\Delta\mu)=z(t,\mu)$. This implies that the derivative $\dfrac{\partial y}{\partial \mu}$ exists and equals $z(t,\mu)$, i.e. it does indeed satisfy equation (2.141) and initial condition (2.142).

Using (2.137), we obtain

$$\frac{dw}{dt}=\frac{f(y(t,\mu+\Delta\mu),t,\mu+\Delta\mu)-f(y(t,\mu),t,\mu+\Delta\mu)}{\Delta\mu}$$

$$+\frac{f(y(t,\mu),t,\mu+\Delta\mu)-f(y(t,\mu),t,\mu)}{\Delta\mu}$$

$$=f_y(y(t,\mu)+\theta_1\Delta y,t,\mu+\Delta\mu)w+f_\mu(y(t,\mu),t,\mu+\theta_2\Delta\mu);$$

$$0\leq\vartheta_1\leq 1,\ \ 0\leq\vartheta_2\leq 1. \tag{2.146}$$

Obviously,

$$w|_{t=t_0}=0. \tag{2.147}$$

Let us write out the equation for the difference $w-z$. Note first that since f_y and f_μ are continuous, and so is $y(t,\mu)$, we have the representation

$$f_y(y(t,\mu)+\theta_1\Delta y,t,\mu+\Delta\mu)=f_y(y(t,\mu),t,\mu)+p(t,\mu,\Delta\mu),$$

where $|p|<\varepsilon_1$ if $|\Delta\mu|<\delta_1(\varepsilon_1)$ uniformly with respect to t. Exactly in the same way,

$$f_\mu(y(t,\mu),t,\mu+\theta_2\Delta\mu)=f_\mu(y(t,\mu),t,\mu)+q(t,\mu,\Delta\mu),$$

where $|q|<\varepsilon_1$ if $|\Delta\mu|<\delta_2(\varepsilon_1)$ uniformly with respect to t. Subtracting (2.143), (2.144) from (2.146), (2.147) we now obtain the following equation for $w-z$

$$\frac{d}{dt}(w-z)=f_y(y(t,\mu),t,\mu)(w-z)+p(w-z)+pz+q \tag{2.148}$$

Here we have

$$(w-z)|_{t=t_0}=0. \tag{2.149}$$

Since z is bounded, we have $|pz+q| < \varepsilon_2$ when $|\Delta\mu| < \delta_3(\varepsilon_2)$. Moreover, since $|f_y| \leq Q$, it follows that $|f_y+p| < Q+\varepsilon_1$. Hence

$$\left|\frac{d}{dt}(w-z)\right| < (Q+\varepsilon_1)|w-z|+\varepsilon_2 \qquad (2.150)$$

and for the estimate of $w-z$ we can again use Lemma 2.1. As a result, we get

$$|w-z| < \frac{\varepsilon_2}{Q+\varepsilon_1}(e^{(Q+\varepsilon_1)H}-1) < \varepsilon \qquad (2.151)$$

when $|\Delta\mu| < \delta(\varepsilon)$ for all $t\in[t_0, t_0+H]$, as required.

Thus we have proved the following theorem.

Theorem 2.11. *Suppose $f(y,t,\mu)$ is defined and continuous with respect to the set of its variables in the domain*

$$D = \{|t-t_0| \leq a, \ |y-y_0| \leq b, \ |\mu-\mu_0| \leq C\}$$

together with its partial derivatives $f_y(y,t,\mu)$ and $f_\mu(y,t,\mu)$. Then the solution $y(t,\mu)$ of the initial value problem (2.133), (2.134) possesses a derivative with respect to μ for every μ in the interval $|\mu-\mu_0| \leq C$; this derivative satisfies equation (2.141) and initial condition (2.142), obtained respectively by taking the derivative of equation (2.133) and initial condition (2.134) with respect to μ.

Equation (2.141) is often referred to as the *equation of variations* with respect to the parameter μ for the equation (2.133).

Remark. The coefficients in the equation of variations (2.141) are continuous with respect to the set of variables t, μ in view of the requirements imposed on f and by Remark 2 following Theorem 2.10. Therefore, applying Theorem 2.10 to the problem (2.141), (2.142) we see that $\dfrac{\partial y}{\partial \mu}$ is continuous with respect to t and with respect to μ.

In an analogous way, we can study the question of the existence of the derivative with respect to the initial value y_0. The result will be that $\dfrac{\partial y}{\partial y_0}$ satisfies the equation of variations

$$\frac{d}{dt}\frac{\partial y}{\partial y_0} = f_y(y(t,y_0),t)\frac{\partial y}{\partial y_0}, \qquad (2.152)$$

obtained by differentiating (2.133) with respect to y_0. There is no inhomogeneous term in this equation, since y_0 does not appear in the right-hand side of (2.133). On the contrary, the initial values for $\dfrac{\partial y}{\partial y_0}$ will be non-zero;

namely, differentiating (2.134) with respect to y_0, we obtain

$$\frac{\partial y}{\partial y_0}\bigg|_{t=t_0} = 1. \tag{2.153}$$

We can also study the existence of derivatives of the solution with respect to the initial value t_0. The derivative $\dfrac{\partial y}{\partial t_0}$ satisfies the equation of variations

$$\frac{d}{dt}\frac{\partial y}{\partial t_0} = f_y(y(t, t_0), t)\frac{\partial y}{\partial t_0}, \tag{2.154}$$

obtained by taking derivatives of equation (2.133) with respect to t_0. In order to obtain the initial conditions, let us replace the problem (2.133), (2.134) by the integral equation

$$y(t, t_0) = y_0 + \int_{t_0}^{t} f(y(\tau, t_0), \tau)d\tau.$$

Taking derivatives with respect to t_0, we get

$$\frac{\partial}{\partial t_0} y(t, t_0) = -f(y(t_0, t_0), t_0) + \int_{t_0}^{t} f_y(y(\tau, t_0), \tau)\frac{\partial y}{\partial t_0} d\tau.$$

Hence, putting $t = t_0$, we obtain

$$\frac{\partial y}{\partial t_0}\bigg|_{t=t_0} = -f(y(t_0, t_0), t_0). \tag{2.155}$$

All these considerations remain valid in the general case (2.127), (2.128) as well, under the condition that the functions f_i in the domain D (see 2.140) possess continuous partial derivatives with respect to all the $y_1, \ldots, y_m$ and to the parameter with respect to which the derivatives are taken. Thus, for example, the derivative $\dfrac{\partial y_i}{\partial \mu_k}$ satisfies the system of equations of variations

$$\frac{d}{dt}\frac{\partial y_i}{\partial \mu_k} = \sum_{l=1}^{m}\frac{\partial f_i}{\partial y_l}\frac{\partial y_l}{\partial \mu_k} + \frac{\partial f_i}{\partial \mu_k},$$

$$\frac{\partial y_i}{\partial \mu_k}\bigg|_{t=t_0} = 0. \tag{2.156}$$

The question of the existence and continuity of the derivatives of higher orders can be studied in a similar way. It may be shown that the existence of continuous partial derivatives of order k of the functions f_i, with respect to $y_1, \ldots, y_m$,

$\mu_k, \ldots, \mu_s$ guarantees the existence of continuous partial derivatives of the k-th order with respect to the parameters $y_{10}, \ldots, y_{m0}, \mu_1, \ldots, \mu_s$ for the solutions of problem (2.127), (2.128). Note that if the functions f_i are analytic functions of their variables, then the solution of problem (2.127), (2.128) turns out to be analytically dependent on the parameters (Poincaré's theorem).

§6. The Method of Successive Approximations (Picard's Method)

When we considered the Cauchy initial value problem for one equation in §2, we proved the existence and uniqueness of the solution by the Euler method, which also turned out to be an effective algorithm for finding the solution numerically. In the present section, we return to the existence and uniqueness of the initial value problem's solution; we shall give its proof by the method of successive approximations, whose main ideas go back to the research work of Picard. Being less efficient from the algorithmic point of view than the Euler method, the method of successive approximations is more general and has wide applications in the study of existence and uniqueness questions in various branches of mathematics. Therefore, knowledge of the main ideas of this method, as shown by the example considered in the present section, is undoubtedly useful.

Consider the initial value problem

$$\frac{dy}{dx} = f(x, y), \quad y(x_0) = y_0, \tag{2.157}$$

where the function $f(x, y)$ is defined and continuous in the rectangle $D = \{|x - x_0| \le a, |y - y_0| \le b\}$. Then there exists a constant M such that

$$|f(x, y)| \le M, \quad (x, y) \in D. \tag{2.158}$$

Moreover, assume that $f(x, y)$ satisfies the Lipschitz condition with respect to y in D

$$|f(x, y_1) - f(x, y_2)| \le N|y_1 - y_2|,$$
$$(x, y_1) \in D, \quad (x, y_2) \in D. \tag{2.159}$$

Let us show that under the conditions imposed on $f(x, y)$, there exists a unique solution of problem (2.157). Our proof will be based on the reduction of problem (2.157) to the equivalent integral equation

$$y(x) = y_0 + \int_{x_0}^{x} f(\xi, y(\xi)) d\xi \tag{2.160}$$

to which the method of successive approximations will be applied. The equivalence of problem (2.157) and this integral equation was established earlier (Lemma 1.1).

We now pass to the construction of the successive approximations. For the zeroth approximation, take an arbitrary continuous function $y_{(0)}(x)$ on the segment $[x_0, X]$ whose graph on $[x_0, X]$, $X = x_0 + H$ is entirely contained in the domain D, and define the successive approximations $y_{(n)}(x)$ by putting

$$y_{(n)}(x) = y_0 + \int_{x_0}^{x} f(\xi, y_{(n-1)}(\xi)) d\xi. \tag{2.161}$$

Just as in § 2, the value of H is determined from the condition $H = \min \{a, b/M\}$. It is easy to prove that under the chosen initial approximation the graphs of the functions $y_{(n)}(x)$ on the closed interval $[x_0, X]$ will also be entirely contained in the domain D. Indeed

$$y_{(1)}(x) = y_0 + \int_{x_0}^{x} f(\xi, y_{(0)}(\xi)) d\xi; \tag{2.162}$$

since the graph of $y_{(0)}(x)$ is contained in the domain D, by the estimate (2.158) we obtain

$$|y_{(1)} - y_0| \le M(x - x_0) \le MH \le b. \tag{2.163}$$

This implies that the graph of the function $y_{(1)}(x)$ on the closed interval $[x_0, X]$ does not leave the domain D. Repeating the arguments carried out above, we shall establish by induction that our statement is correct for any of the approximations.

Lemma 2.4. *If the function $f(x, y)$ is continuous in the domain D and satisfies the Lipschitz condition (2.159), then the sequence $\{y_{(n)}(x)\}$ constructed according to formula (2.161) converges uniformly on $[x_0, X]$.*

Proof. Consider the series

$$S(x) = y_{(0)}(x) + [y_{(1)}(x) - y_{(0)}(x)] + \ldots + [y_{(n)}(x) - y_{(n-1)}(x)] + \ldots \tag{2.164}$$

It is obvious that the n-th partial sum $S_n(x)$ of the series is equal to the n-th term of the sequence $\{y_{(n)}(x)\}$. Let us estimate the terms of the series. Obviously

$$|y_{(1)}(x) - y_{(0)}(x)| \le |y_{(1)}(x) - y_0| + |y_{(0)}(x) - y_0| \le 2b,$$

$$|y_{(2)}(x) - y_{(1)}(x)| \le \int_{x_0}^{x} |f(\xi, y_{(1)}(\xi)) - f(\xi, y_{(0)}(\xi))| d\xi.$$

Then by condition (2.159) we have the estimate

$$|y_{(2)}(x) - y_{(1)}(x)| \leq N \int_{x_0}^{x} |y_{(1)}(\xi) - y_{(0)}(\xi)| d\xi \leq 2bN(x - x_0).$$

Similarly

$$|y_{(3)}(x) - y_{(2)}(x)| \leq N \int_{x_0}^{x} |y_{(2)}(\xi) - y_{(1)}(\xi)| d\xi$$

$$\leq 2bN^2 \int_{x_0}^{x} (\xi - x_0) d\xi = 2bN^2 \frac{(x - x_0)^2}{2!}. \qquad (2.165)$$

Arguing by induction, we obtain the following estimate for the n-th term

$$|y_{(n)}(x) - y_{(n-1)}(x)| \leq 2bN^{n-1} \frac{(x - x_0)^{n-1}}{(n-1)!}. \qquad (2.166)$$

The estimate (2.166) implies that the terms of the functional series on our closed interval are bounded from above by the terms of the convergént numerical series

$$|y_{(n)}(x) - y_{(n-1)}(x)| \leq 2bN^{n-1} \frac{H^{n-1}}{(n-1)!}, \qquad (2.167)$$

which is a sufficient condition for the uniform convergence of the series (2.164). The lemma is proved.

The lemma proved above allows us to establish the following theorem.

Theorem 2.12 *(existence). If the function $f(x, y)$ is continuous in D and satisfies the Lipschitz condition (2.159), then there exists a solution of the initial value problem (2.157) on the closed interval $[x_0, X]$.*

Proof. In view of lemma 1.1, it is sufficient to prove the existence of a solution of the integral equation (2.160) on the closed interval $[x_0, X]$. According to Lemma 2.4, the sequence $\{y_{(n)}(x)\}$ constructed by means of formula (2.161) converges uniformly on $[x_0, X]$. Since all the terms of the sequence $\{y_{(n)}(x)\}$ are continuous functions by construction, it follows that the limiting function $y(x)$ is continuous on $[x_0, X]$. The fact that the sequence $\{y_{(n)}\}$ converges uniformly on $[x_0, X]$ is a sufficient condition for passage to the limit in formula (2.161). Passing to the limit as $n \to \infty$, we see that the limiting function of the sequence

$$y(x) = \lim_{n \to \infty} y_{(n)}(x) \qquad (2.168)$$

satisfies the integral equation (2.160), but this is equivalent to the original problem (2.157). The theorem is proved.

Now let us discuss the uniqueness of the solution.

Theorem 2.13 *(uniqueness). The integral equation (2.160) possesses no more than one continuous solution on* $[x_0, X]$.

Proof. Assume that equation (2.160) possesses two different solutions $y_1(x)$ and $y_2(x)$ on $[x_0, X]$ and consider their difference

$$z(x) = y_1(x) - y_2(x). \tag{2.169}$$

Obviously,

$$z(x) = \int_{x_0}^{x} \{f(\xi, y_1(\xi)) - f(\xi, y_2(\xi))\} d\xi, \quad x \in [x_0, X]. \tag{2.170}$$

First we shall consider (2.170) on the closed interval $[x_0, x_1]$; the value of x_1 will be chosen later.

Using the Lipschitz condition (2.159), we get

$$|z(x)| \leq N(x - x_0) \sup_{[x_0, x]} |y_1(\xi) - y_2(\xi)|$$
$$\leq N(x_1 - x_0) \sup_{[x_0, x_1]} |z(\xi)|, \quad x \in [x_0, x_1]. \tag{2.171}$$

Hence

$$\sup_{[x_0, x_1]} |z(x)| \leq N(x_1 - x_0) \sup_{[x_0, x_1]} |z(x)|. \tag{2.172}$$

Now choosing x_1 so as to have $N(x_1 - x_0) < 1$, we see that (2.172) is possible only under the condition that

$$\sup_{[x_0, x_1]} |z(x)| = 0 \quad \text{i.e.} \quad z(x) \equiv 0 \quad \text{for} \quad x \in [x_0, x_1]$$

and (2.170) may be written in the form

$$z(x) = \int_{x_1}^{x} \{f(\xi, y_1(\xi)) - f(\xi, y_2(\xi))\} d\xi, \quad x \in [x_1, X]. \tag{2.173}$$

Repeating the argument above the number of times required, we see that $z(x) \equiv 0$ for $x \in [x_0, X]$, which proves the theorem.

Some remarks concerning the theorems that we have just proved are called for.

Remarks. 1. We have proved the existence theorem by showing that, for any zeroth approximation $y_{(0)}(x)$ which is a continuous function on $[x_0, X]$ whose graph does not leave the domain D, the sequence $\{y_{(n)}\}$ converges to the solution of the given problem. In many cases it is convenient to choose the initial value y_0 for the zeroth approximation $y_0(x)$, putting $y_{(0)}(x) \equiv y_0$.

2. The method of successive approximations may be used not only to prove existence, but also to construct the solution of specific problems. Then its efficiency depends both on the class of functions $f(x, y)$ for which effective algorithms computing the right-hand side of formula (2.161) exist, and on the choice of the initial approximation.

3. We have considered the application of the successive approximation method to prove the existence and uniqueness of initial value problem solutions in the case of one scalar first order equation. Similar considerations are valid in the case of initial value problems for a normal system.

§7. The Contraction Mapping Theorem

The method of successive approximations considered in the previous section is based on a general mathematical principle, known as the "contraction mapping theorem", whose main ideas will be developed in the present section.

We shall consider an arbitrary complete metric space M. Recall that the space M is a *metric space* if to every pair of its elements x, y a number $\varrho(x, y)$ is assigned so that the following conditions hold:

1) $\varrho(x, y) \geq 0$ and $\varrho(x, y) = 0$ only if $x = y$;
2) $\varrho(x, y) = \varrho(y, x)$
3) for all $x, y, z \in M$ we have the triangle inequality

$$\varrho(x, y) \leq \varrho(x, z) + \varrho(y, z).$$

The number $\varrho(x, y)$ is usually called the *distance* between the elements x and y. A metric space M is said to be *complete* if every Cauchy sequence of elements $\{x_m\}$ of this space[2] converges to some element $x \in M$, i.e. $\exists x \in M$ such that $\lim\limits_{m \to \infty} \varrho(x, x_m) = 0$ (we denote $x = \lim\limits_{m \to \infty} x_m$).

A well-known example of a complete metric space is the space of functions $y(x)$, continuous on the closed interval $x \in [a, b]$, when the distance $\varrho(y, z)$ between elements $y(x)$ and $z(x)$ is given in the form

$$\varrho(y, z) = \sup_{x \in [a, b]} |y(x) - z(x)|. \qquad (2.174)$$

The contraction mapping theorem is a powerful method used to study existence and uniqueness for functional equations in metric spaces.

---

[2] The sequence $\{x_m\}$ is called Cauchy if the Cauchy criterion holds for it: $\lim\limits_{\substack{n \to \infty \\ m > 0}} \varrho(x_n, x_{n+m}) = 0.$

Suppose that in the complete metric space M, we are given the operator A possessing the following properties:

a) the operator A maps the space M into itself, i.e. it sends points of the space M into points of the same space:

$$\forall x \in M, \quad Ax = y, \quad y \in M; \qquad (2.175)$$

b) the operator A brings all elements of the space M nearer to each other, i.e. for any pair x_1, x_2 of elements of the space M we have the inequality

$$\varrho(Ax_1, Ax_2) \leq \alpha \varrho(x_1, x_2), \qquad (2.176)$$

where $\alpha < 1$.

We have the following

Theorem 2.14 *(the contraction mapping theorem). If, in a complete metric space M, we are given an operator A satisfying conditions a) and b), then the functional equation*

$$Ax = x \qquad (2.177)$$

possesses a unique solution in the space M.

Proof. Choose an arbitrary element $x_0 \in M$ and construct the sequence $\{x_n\}$:

$$x_n = Ax_{n-1}. \qquad (2.178)$$

Let us prove that the sequence thus constructed is Cauchy. Indeed,

$$\varrho(x_{n+1}, x_n) = \varrho(Ax_n, Ax_{n-1}) \leq \alpha \varrho(x_n, x_{n-1})$$

$$= \alpha \varrho(Ax_{n-1}, Ax_{n-2}) \leq \ldots \leq \alpha^n \varrho(x_1, x_0). \qquad (2.179)$$

Using this estimate and the triangle inequality, we get

$$\varrho(x_{n+m}, x_n) \leq \varrho(x_{n+m}, x_{n+m-1}) + \varrho(x_{n+m-1}, x_{n+m-2})$$

$$+ \ldots + \varrho(x_{n+1}, x_n) \leq \alpha^n \varrho(x_1, x_0) \, (\alpha^{m-1} + \ldots + 1)$$

$$= \alpha^n \frac{1 - \alpha^m}{1 - \alpha} \, \varrho(x_1, x_0) \leq \frac{\alpha^n}{1 - \alpha} \, \varrho(x_1, x_0), \qquad (2.180)$$

which under the condition $\alpha < 1$ proves that the sequence $\{x_n\}$ is Cauchy. Since the space M is complete, it follows that the sequence $\{x_n\}$ converges, i.e. there exists an element $x \in M$ such that

$$x = \lim_{n \to \infty} x_n, \qquad (2.181)$$

Now consider the element $y = Ax$; let us prove that it coincides with the element x. Indeed

$$\varrho(x, y) \le \varrho(x, x_{n+1}) + \varrho(x_{n+1}, y), \qquad (2.182)$$

where x_{n+1} is an arbitrary element of the sequence $\{x_n\}$. Let us estimate the last summand

$$\varrho(x_{n+1}, y) = \varrho(Ax_n, Ax) \le \alpha \varrho(x_n, x). \qquad (2.183)$$

In view of the convergence of the sequence $\{x_n\}$ (proved above), for any $\varepsilon > 0$ we can indicate an N such that $n > N$ implies $\varrho(x_n, x) < \varepsilon/2$.

Then by (2.182) if follows that

$$\varrho(x, y) < \varepsilon, \qquad (2.184)$$

hence, since ε was arbitrary, we obtain $\varrho(x, y) = 0$, i.e. $x = y$ and therefore the solution of equation (2.177) exists.

The proof of the uniqueness of the solution of equation (2.177) shall be carried out by *reductio ad absurdum*. Suppose x and $\bar{x}$ are two different solutions of this equation

$$Ax = x, \quad A\bar{x} = \bar{x}. \qquad (2.185)$$

Consider

$$\varrho(x, \bar{x}) = \varrho(Ax, A\bar{x}) \le \alpha \varrho(x, \bar{x}). \qquad (2.186)$$

Then for $\alpha < 1$ the inequality (2.186) is possible only if $\varrho(x, \bar{x}) = 0$, i.e. if $\bar{x} = x$. Thus the theorem is entirely proved.

The theorem proved above has a simple geometric interpretation. If we consider the elements of the metric space as points of a certain set, the contracting property b) of the operator A means that the distance $\varrho(x_1, x_2)$ between the images $y_1 = Ax_1$, $y_2 = Ax_2$ of the points x_1 and x_2 is less than the distance $\varrho(x_1, x_2)$ between the given points x_1 and x_2. Therefore, when we construct the sequence $\{x_n\}$ according to formula (2.178), the distance between neighbouring points decreases when the number n increases unboundedly and in the limit we obtain a fixed point x which the operator A sends into itself, $Ax = x$.

Let us apply the theorem on the contraction mapping theorem to prove the existence and uniqueness of the solution of the following initial value problem

$$\frac{dy}{dx} = f(x, y), \quad y(x_0) = y_0, \qquad (2.187)$$

which, as we have established in the previous section, is equivalent to the integral equation

$$y(x) = y_0 + \int_{x_0}^{x} f(\xi, y(\xi)) d\xi. \qquad (2.188)$$

Consider the operator

$$Ay \equiv y_0 + \int\limits_{x_0}^{x} f(\xi, y(\xi)) d\xi \qquad (2.189)$$

in the complete metric space M of functions $y(x)$ continuous on the closed interval $[x_0, X]$. Let us show that the operator A satisfies conditions a) and b). In the previous section, we showed that if the function $f(x, y)$ is continuous in the rectangle $D = \{|x - x_0| \le a, |y - y_0| \le b\}$ and satisfies the Lipschitz condition with respect to y in D, then the application of the operator A to a continuous function $y(x)$ on $[x_0, X]$ whose graph does not leave D also yields a continuous function $Ay(x)$ whose graph does not leave D. Thus the operator A satisfies condition a). It remains for us to prove the contracting property of operator A. To do this consider

$$\varrho(Ay_1, Ay_2) = \sup\limits_{x \in [x_0, X]} \left| \int\limits_{x_0}^{x} f(\xi, y_1(\xi)) d\xi - \int\limits_{x_0}^{x} f(\xi, y_2(\xi)) d\xi \right|$$

$$\le \sup\limits_{x \in [x_0, X]} \int\limits_{x_0}^{x} |f(\xi, y_1(\xi)) - f(\xi, y_2(\xi))| d\xi. \qquad (2.190)$$

Using the Lipschitz condition for the function $f(x, y)$, we obtain

$$\varrho(Ay_1, Ay_2) \le N \sup\limits_{x \in [x_0, X]} \int\limits_{x_0}^{x} |y_1(\xi) - y_2(\xi)| d\xi$$

$$\le N|x - x_0| \sup\limits_{x \in [x_0, X]} |y_1(x) - y_2(x)| \le N H \varrho(y_1, y_2). \qquad (2.191)$$

Now choose H so as to have

$$NH = \alpha < 1. \qquad (2.192)$$

Then the operator A will be contracting and, by Theorem 2.14, we can claim the existence and uniqueness of the solution of the initial value problem (2.187) on the closed interval $[x_0, X]$. The extension of the solution to a larger interval can be carried out by the methods described above.

Remark. The contraction mapping theorem was applied to prove the existence and uniqueness of the solution of the initial value problem (2.187) for one scalar equation. By using the contraction mapping theorem, it is easy to prove a similar theorem in the case of a normal system.

Chapter III
Linear Differential Equations

§ 1. The Pendulum Equation as an Example of a Linear Equation. The Main Properties of Linear Equations with Constant Coefficients

A linear differential equation of the n-th order is an equation of the form

$$a_0(x) y^{(n)} + a_1(x) y^{(n-1)} + \ldots + a_n(x) y = f(x). \tag{3.1}$$

This equation possesses a series of remarkable properties which facilitate its study and, in many cases, its solution. The study of these properties is the topic of the present chapter.

In the applications, linear equations are obtained in a natural way if one neglects higher order terms (see §2, Chapter 1).

We shall familiarize ourselves with the main properties of linear equations using the example of the pendulum equation (see (1.26) Sect. 2.2 in Chapter 1),

$$y'' + \alpha y' + ky = f(t), \tag{3.2}$$

which is a linear second order equation with constant coefficients.

First consider the case $f = 0$. In this case the equation is known as *homogeneous*. From the physical point of view this means that the pendulum moves freely and no external forces act upon it

$$y'' + \alpha y' + ky = 0. \tag{3.3}$$

We shall look for the solution of this equation in the form $y = e^{\lambda t}$, where λ is an unspecified constant. Substituting the required form of the solution in (3.3) and dividing by $e^{\lambda t}$, we get

$$\lambda^2 + \alpha\lambda + k = 0. \tag{3.4}$$

This equation is known as the *characteristic equation* of the differential equation (3.3). It must be satisfied by λ in order that $e^{\lambda t}$ be a solution of (3.3). Solving equation (3.4), we obtain

$$\lambda_{1,2} = \frac{-\alpha \pm \sqrt{\alpha^2 - 4k}}{2}.$$

Consider several cases:

a). $\alpha^2 - 4k > 0$. Physically this corresponds to sufficiently strong friction (resistance) of the medium. Both roots λ_1 and λ_2 in this case are real, distinct and negative and we obtain two corresponding solutions $y_{(1)} = e^{\lambda_1 t}, y_{(2)} = e^{\lambda_2 t}$.

Consider the initial value problem

$$y(0) = y_0^0, \quad y'(0) = y_1^0. \tag{3.5}$$

For any two functions $y_1(x)$, $y_2(x)$ which are differentiable n times, we have the following identity

$$(C_1 y_1(x) + C_2 y_2(x))^{(n)} = C_1 y_1^{(n)} + C_2 y_2^{(n)}. \tag{3.6}$$

where C_1 and C_2 are constants. Using this identity, it is easy to verify that the expression

$$y = C_1 y_{(1)} + C_2 y_{(2)} = C_1 e^{\lambda_1 t} + C_2 e^{\lambda_2 t}, \tag{3.7}$$

where C_1 and C_2 are arbitrary constants (a linear combination of $y_{(1)}$ and $y_{(2)}$) is a solution of equation (3.3). These constants may be determined uniquely from the initial conditions (3.5). Indeed, if we substitute (3.7) into (3.5), we get

$$y_0^0 = C_1 + C_2, \quad y_1^0 = C_1 \lambda_1 + C_2 \lambda_2.$$

Since $\lambda_1 \neq \lambda_2$, the determinant of this linear algebraic system with respect to C_1 and C_2 does not vanish. The solution of the initial value problem

$$y = \frac{\lambda_2 y_0^0 - y_1^0}{\lambda_2 - \lambda_1} e^{\lambda_1 t} + \frac{\lambda_1 y_0^0 - y_1^0}{\lambda_1 - \lambda_2} e^{\lambda_2 t}, \tag{3.8}$$

obtained in this way does not oscillate but approaches the state of equilibrium $y = 0$ when t increases.

Since any known solution of Eq. (3.3) satisfies a certain initial condition (3.5) and, since the initial condition (3.5) uniquely determines the solution (3.8), we can say that formula (3.7) contains any solution of equation (3.3). On the other hand, for any values of the constants, formula (3.7) gives a solution of equation (3.3). Therefore, formula (3.7) contains all the solutions of Eq. (3.3) and only solutions of this equation. Any formula possessing this property will be called the *general solution. Formula (3.7) is the general solution of equation (3.3).*

b) $\alpha^2 - 4k < 0$. Physically this corresponds to sufficiently weak friction (resistance) of the medium. In this case λ_1 and λ_2 are complex conjugate: $\lambda_2 = \lambda_1^*$ and

$$y_{(1)} = e^{\lambda_1 t} = e^{-\alpha t/2} \left(\cos \frac{\beta}{2} t + i \sin \frac{\beta}{2} t \right), \qquad y_{(2)} = y_{(1)}^*,$$

where $\beta = \sqrt{4k - \alpha^2}$.

Using identity (3.6), it is easy to see that $y_1 = \operatorname{Re} y_{(1)}$, $y_2 = \operatorname{Im} y_{(1)}$ are also solutions of equation (3.2). Indeed,

$$(y_1 + iy_2)'' + \alpha(y_1 + iy_2)' + k(y_1 + iy_2)$$
$$= (y_1'' + \alpha y_1' + ky_1) + i(y_2'' + \alpha y_2' + ky_2) = 0.$$

hence by setting the real and imaginary parts equal to each other, we obtain the required result. Take a linear combination of y_1 and y_2

$$y = C_1 y_1 + C_2 y_2 = C_1 e^{-\alpha t/2} \cos \frac{\beta}{2} t + C_2 e^{-\alpha t/2} \sin \frac{\beta}{2} t. \qquad (3.9)$$

Then, as before, C_1 and C_2 are uniquely determined by conditions (3.5) and therefore (3.9) is the general solution of equation (3.3). Note that in the case considered we can still take (3.7) for the general solution, but now the constants C_1 and C_2 will be complex.

The solution of problem (3.5):

$$y = y_0^0 e^{-\alpha t/2} \cos \frac{\beta}{2} t + \frac{2}{\beta} \left(y_1^0 + \frac{\alpha}{2} y_0^0 \right) e^{-\alpha t/2} \sin \frac{\beta}{2} t \qquad (3.10)$$

describes an oscillating process. The oscillations are damped according to the law $\exp[-\alpha t/2]$. When t increases, the solution also tends to the equilibrium position $y = 0$.

If $\alpha = 0$ (there is no resistance), we obtain periodic oscillations with frequency $\omega_0 = \sqrt{k}$

$$y = y_0^0 \cos \omega_0 t + \frac{1}{\omega_0} y_1^0 \sin \omega_0 t. \qquad (3.11)$$

c) $\alpha^2 - 4k = 0$. In this case the method described above yields only one solution $y_{(1)} = e^{\lambda t}$, where $\lambda = -\alpha/2$. However, it is easy to check directly that in this case $y_{(2)} = te^{\lambda t}$ is also a solution. Taking the linear combination of these two solutions, we can satisfy condition (3.5). In practice λ_1 and λ_2 are never exactly equal to each other, so that such a solution describes a mathematical abstraction corresponding to the case when λ_1 and λ_2 are close to each other.

Now let us consider forced oscillations under the action of a periodic external force. They are described by the equation (3.2), where $f = A \cos \omega t$ (A, ω are constants). Along with this equation, we consider the following equation for the complex unknown function z:

$$z'' + \alpha z' + kz = Ae^{i\omega t}. \qquad (3.12)$$

Substituting the expression $z = \tilde{y}_1 + i\tilde{y}_2$ into this equation and setting the real and imaginary parts equal to each other, we see that $\tilde{y}_1$ satisfies Eq. (3.2) in which $f = A \cos \omega t$ while $\tilde{y}_2$ satisfies Eq. (3.2) in which $f = A \sin \omega t$. Thus, to obtain the required solution of Eq. (3.2), we must find a solution of Eq. (3.12) and take its real part.

It is natural lo look for the solution of Eq. (3.12) in the form

$$z = ae^{i\omega t}, \qquad (3.13)$$

where a is an unspecified constant. Substituting (3.13) into (3.12) and dividing by $e^{i\omega t}$, we obtain $a = A/(-\omega^2 + i\alpha\omega + k)$ and therefore

$$\tilde{y}_1 = A \frac{k - \omega^2}{(k - \omega^2)^2 + \alpha^2 \omega^2} \cos \omega t + A \frac{\alpha\omega}{(k - \omega^2)^2 + \alpha^2 \omega^2} \sin \omega t. \qquad (3.14)$$

(3.14) is a particular solution of equation (3.2) in which $f = A \cos \omega t$; this solution is periodic with frequency equalling the frequency ω of the external force. This solution, however, does not satisfy (3.5). Let us add a linear combination of solutions of the homogeneous equation (3.3) (where for the sake of brevity we only consider the case $\alpha^2 - 4k < 0$):

$$y = \tilde{y}_1 + C_1 y_1 + C_2 y_2. \qquad (3.15)$$

Using (3.6), we see that this expression is a solution of the same non-homogeneous equation (3.2) and, using the fact that the constants C_1 and C_2 are arbitrary, we can choose them in such a way as to satisfy (3.5). Indeed, C_1 and C_2 can be found from the algebraic system of equations used to obtain (3.10), but with different right hand sides. The solution satisfying (3.5) will be of the form

$$y = \tilde{y}_1 + [y_0^0 - \tilde{y}_1(0)]e^{-\alpha t/2} \cos \frac{\beta}{2} t$$
$$+ \frac{2}{\beta} \left[y_1^0 - \tilde{y}_1'(0) + \frac{\alpha}{2} (y_0^0 - \tilde{y}_1(0)) \right] e^{-\alpha t/2} \sin \frac{\beta}{2} t, \qquad (3.16)$$

while (3.15) turns out to be the general solution of the non-homogeneous equation (3.2), where $f = A \cos \omega t$. It is clear from (3.15) that *the general solution of the non-homogeneous equation is the sum of a particular solution of the same non-homogeneous equation and the general solution of the corresponding homogeneous equation.*

When t increases, all the terms in formula (3.16) except the first one are damped and only the forced oscillations $\tilde{y}_1$ remain.

Note the important phenomenon called resonance. The solution $\tilde{y}_1$ loses its meaning if there is no friction in the given system ($\alpha = 0$) and the frequency ω of the external force is equal to the frequency $\omega_0 = \sqrt{k}$ with which the pendulum oscillates in the absence of external force (see (3.14)), since in this case the denominator vanishes.

In order to find a particular solution in this case, i.e. a particular solution of the equation

$$y'' + ky = A \cos \omega_0 t, \tag{3.17}$$

let us once again consider the complex form

$$z'' + kz = A e^{i\omega_0 t}. \tag{3.18}$$

Note that the roots of the characteristic equation are equal to $\lambda_{1,2} = \pm i\omega_0$. Let us try to find z in the form

$$z = at e^{i\omega_0 t}. \tag{3.19}$$

Substituting (3.19) into (3.18), we can determine a and obtain $z = A \dfrac{t}{2 i\omega_0} e^{i\omega_0 t}$. Re z gives the required particular solution of Eq. (3.17):

$$\tilde{y}_1 = A \frac{t}{2 \omega_0} \sin \omega_0 t. \tag{3.20}$$

Since the complete absence of friction and the exact equality between ω and ω_0 are never realized, no solutions of this type are met with in practice. We actually meet with (3.14), but, if the frequency ω is close to ω_0, while α is small, then the denominator in (3.14) is small and the amplitude of the solution is very large. Thus physically the phenomenon of resonance consists in a notable increase of the amplitude of forced oscillations (3.14) in the case when $\omega \sim \omega_0$ and α is small.

From the mathematical point of view, resonance is the case when we have $f(t) = S(t) e^{\kappa t}$ in (3.2), where $S(t)$ is a polynomial, while κ coincides with the root of the characteristic equation. In the case of Eq. (3.18) considered above, $\kappa = i\omega_0$, i.e. κ coincides with one of the roots of the characteristic equation.

Thus, using a second-order equation as an example, we have shown a series of typical properties of linear equations with constant coefficients. It turns out that these properties are general in character. For the moment, we shall state them without proof, for equations of order n, as natural generalizations of what we observed for equations of the second order. The proofs shall be given later (in § 5).

First consider the homogeneous equation

$$a_0 y^{(n)} + a_1 y^{(n-1)} + \ldots + a_n y = 0 \qquad (a_i = \text{const}). \tag{3.21}$$

To (3.21) assign its characteristic equation (compare with (3.4))

$$a_0\lambda^n + a_1\lambda^{n-1} + \ldots + a_n = 0. \tag{3.22}$$

This is an algebraic equation of order n and has the roots $\lambda_k = p_k + iq_k$ ($k = 1, \ldots, n$).

1. If all the λ_k are real and distinct, then taking the linear combination

$$y = \sum_{k=1}^{n} C_k y_{(k)}, \quad \text{where} \quad y_{(k)} = e^{\lambda_k t}, \tag{3.23}$$

we can obtain any solution of Eq. (3.21) by determining the constants $C_1, \ldots, C_n$ from the initial conditions

$$y(t_0) = y_1^0, \quad y'(t_0) = y_2^0, \ldots, y^{(n-1)}(t_0) = y_n^0 \tag{3.24}$$

(compare with (3.7), (3.5), i.e. formula (3.23) is the general solution of equation (3.21)).

2. If some of the λ_k are complex, then statement 1 remains valid, but the constants C_k determined from (3.24) will be complex numbers and the solution will be presented in complex form. In order to obtain the solution in real form, instead of the pair of solutions $y = e^{(p+iq)t}$ and $y^* = e^{(p-iq)t}$ corresponding to two complex conjugate roots $\lambda = p \pm iq$ (since the characteristic euqation has real coefficients, it follows that, together with the root $\lambda = p + iq$, we shall also have the root $\lambda^* = p - iq$), we may choose, from our family of solutions, the pair of real solutions $\operatorname{Re} y = e^{pt} \cos qt$ and $\operatorname{Im} y = e^{pt} \sin qt$ (compare with 3.9)).

3. If λ is a multiple root (of the characteristic equation (3.22)) of multiplicity m, then to this root correspond m solutions $e^{\lambda t}, te^{\lambda t}, \ldots, t^{m-1}e^{\lambda t}$ (the generalization of case c), where $m = 2$).

Bringing all this together, we can state the following general rule:

Suppose the characteristic equation (3.22) possesses r real roots λ_k of multiplicity m_k and the other roots are complex conjugate of the form $\lambda_l = p_l \pm iq_l$ and of multiplicity m_l. Then the general solution of equation (3.21) may be written in the form

$$y = \sum_{k=1}^{r} R_k(t) e^{\lambda_k t} + \sum_{l=1}^{\frac{n-r}{2}} (P_l(t) e^{p_l t} \cos q_l t + Q_l(t) e^{p_l t} \sin q_l t), \tag{3.25}$$

where $R_k(t)$, $P_l(t)$, $Q_l(t)$ are polynomials of degrees $m_k - 1$, $m_l - 1$, $m_l - 1$ respectively, whose coefficients are arbitrary. These coefficients are uniquely determined by the initial conditions (3.24).

In a similar way, generalizing the facts obtained for second order equations, it is possible to state the rule for constructing the particular and general solution of the non-homogeneous equation

$$a_0 y^{(n)} + a_1 y^{(n-1)} + \ldots + a_n y = S(t) e^{\kappa t}, \tag{3.26}$$

where $S(t)$ is a polynomial of degree s and κ is a constant which may be complex in general.

Suppose that in Eq. (3.26) κ does not coincide with any of the roots λ_k of the characteristic equation (3.22) (the so-called non-resonance case). Then a particular solution of Eq. (3.26) may be written in the form

$$y = T(t)e^{\kappa t}, \tag{3.27}$$

where $T(t)$ is a polynomial of the same degree as $S(t)$. The coefficients of the polynomial $T(t)$ may be determined from the algebraic equations obtained by substituting (3.27) into (3.26) and setting terms in the same power of t equal to each other (compare (3.12), (3.13); in that simplest case $S(t)$ was the constant A, i.e. a polynomial of degree zero, while the polynomial $T(t)$ was also a constant).

If κ coincides with the root of characteristic equation λ of multiplicity m (the so-called resonance case), then a particular solution of (3.26) may be found in the form

$$y = T(t)t^{m}e^{\kappa t}, \tag{3.28}$$

where $T(t)$ is a polynomial of the same degree as $S(t)$. The coefficients of $T(t)$ can still be determined by substituting (3.28) into equation (3.26) (compare with (3.19), where the factor t appeared in connection with the multiplicity $m=1$ of the root $i\omega_0$).

If $\alpha + i\beta = \kappa$ is a complex number, then the real (respectively the imaginary) part of the solution (3.28) is the solution of the equation with right-hand side $S(t)e^{\alpha t} \cos \beta t$ (respectively $S(t)e^{\alpha t} \sin \beta t$).

The general solution of the non-homogeneous equation (3.26) can be represented as the sum of the general solution (3.25) of the homogeneous equation (3.21) and the particular solution (3.27) or (3.28) of the non-homogeneous equation (3.26) (compare with (3.15)).

All these statements will be proved rigorously later on (see theorems 3.12, 3.13, 3.14, 3.15).

§2. General Properties of n-th Order Equations

Now consider equation (3.1). If in the domain where the independent variable varies we have $a_0(x) \neq 0$, then, dividing by $a_0(x)$ and denoting the coefficients and the right-hand side thus obtained by the same symbols $a_1(x), \ldots, a_n(x), f(x)$, we get

$$y^{(n)} + a_1(x)y^{(n-1)} + \ldots + a_n(x)y = f(x). \tag{3.29}$$

Definition. Equation (3.29) is called *homogeneous* if $f(x) \equiv 0$ and non-homogeneous in the converse case.

Suppose the $a_i(x)$ $(i=1,\ldots,n)$ are continuous on some interval X (X can be a finite interval as well as an infinite one, say $(-\infty, \infty)$). The general existence and uniqueness theorem (see §4 in Chapter 2) guarantees that on a certain closed interval $|x-x_0|\leq H$ belonging to X there exists a unique solution $y(x)$ of Eq. (3.29) satisfying the initial condition

$$y(x_0)=y_1^0, \quad y'(x_0)=y_2^0,\ldots,y^{(n-1)}(x_0)=y_n^0. \tag{3.30}$$

For equation (3.29) a stronger statement may be proved.

Theorem 3.1. *If $a_i(x), f_i(x)$ $(i=1,\ldots,n)$ are continuous on X, then the solution of the initial value problem (3.29), (3.30) exists and is unique everywhere on X.*

Since the initial value problem for the n-th order equation is a particular case of the initial value problem for a system of n equations of the first order (see §4 in Chapter 2) it is sufficient to obtain Theorem 3.1 as a particular case of a similar statement for systems of linear equations of the form

$$\frac{dy_i}{dt} = \sum_{k=1}^{n} a_{ik}(x)\,y_k + f_i(x) \quad (i=1,\ldots,n), \tag{3.31}$$

with the corresponding initial conditions

$$y_i(x_0)=y_i^0 \quad (i=1,\ldots,n). \tag{3.32}$$

Theorem 3.2. *If $a_{ik}(x)$, $f_i(x)$ $(i,k=1,\ldots,n)$ are continuous on X, then the solution of problem (3.31), (3.31) exists and is unique on X.*

It is sufficient to prove that the solution exists and is unique on any closed interval $[x_0, x_0 + \Delta]\subset X$. The existence and uniqueness theorem of §4 in Chapter 2 guarantees the solution on a certain closed interval $[x_0, x_0+H]$, as was indicated above. The point $(x_0 + H, y_i(x_0 + H))$ may be taken to be a new initial point (see Remark 2 concerning the theorems in §2), giving a solution on a larger closed interval $[x_0, x_0 + H_1]$, $H_1 > H$, etc. Suppose $[x_0, x_0 + \bar{H})$, where $\bar{H}\leq \Delta$ is the maximal semi-interval on which a unique solution of problem (3.31), (3.32) exists. Choose an arbitrary sequence $H_l \to \bar{H}$. Let us check that the limit $\lim_{l\to\infty} y_i(x_0 + H_l)$ exists. Suppose

$$N=\max_{i,k}\ \sup_{[x_0,x_0+\Delta]} |a_{ik}(x)|, \quad \gamma=\max_i\ \sup_{[x_0,x_0+\Delta]} |f_i(x)|$$

Then on any closed interval we have the inequality

$$\left|\frac{dy_i}{dx}\right|\leq N \sum_{j=1}^{n} |y_j(x)|+\gamma.$$

Introducing $\varrho(x) = \sqrt{\sum\limits_{i=1}^{n} y_i^2(x)}$ and repeating the same arguments as in the proof

of Lemma 2.3 in §4 of Chapter 2, we obtain

$$\varrho(x) \le \varrho(x_0) e^{Nn^2\bar{H}} + \frac{\gamma}{Nn} (e^{Nn^2\bar{H}} - 1) = B,$$

and therefore for any $[x_0, x_0 + H_l]$ we have the inequality

$$|y_i(x)| \le \varrho(x_0) e^{Nn^2\bar{H}} + \frac{\gamma}{Nn} (e^{Nn^2\bar{H}} - 1),$$

so that we get the inequality

$$\left|\frac{dy_i}{dx}\right| \le nNB + \gamma = C.$$

Using it and the fact that $H_l \to \bar{H}$, we obtain

$$|y_i(x_0 + H_{l+m}) - y_i(x_0 + H_l)| \le C|H_{l+m} - H_l| < \varepsilon$$

for $n > N(\varepsilon)$ and any m. Hence, according to the Cauchy criterion, we can conclude that the sequence $y_i(x_0 + H_l)$ converges to some limit $\bar{y}_i$. We shall view this limit as the value of y_i at the point $x_0 + \bar{H}$, i.e. we put $y_i(x_0 + \bar{H}) = \bar{y}_i$. Thus the integral curve turns out to be continuously extended up to the point $x_0 + \bar{H}$.

According to the same system of equations (3.31), the derivatives $\dfrac{dy_i}{dx}$ possess the same property. Therefore, in the case $\bar{H} = \varDelta$, the theorem is proved. Consider the case $\bar{H} < \varDelta$. It is easy to see that this case cannot arise. Indeed, taking $x_0 + \bar{H}$, $y_i(x_0 + \bar{H})$ to be the new initial point, we can extend the solution to the interval $[x_0, x_0 + \bar{H} + \delta]$, $\delta > 0$. This contradicts the definition of $\bar{H}$.

Thus $\bar{H} = \varDelta$, i.e. the solution exists and is unique on the closed interval $[x_0, x_0 + \varDelta]$, as required.

We shall put off further study of solution (3.31) until §6 and return to (3.29). For equation (3.29) we have the following theorem, known as the superposition principle.

Theorem 3.3. *Suppose that the right-hand side $f(x)$ of equation (3.29) is a linear combination of the functions $f_i(x)$, i.e. $f(x) = \sum\limits_{i=1}^{k} \alpha_i f_i(x)$, where the α_i are constants; let $y_i(x)$ be solutions of the equations*

$$y_i^{(n)} + a_1(x) y_i^{(n-1)} + \ldots + a_n(x) y_i = f_i(x). \tag{3.33}$$

Then the linear combination of the $y_i(x)$ with the same coefficient α_i, i.e. the function $y(x) = \sum\limits_{i=1}^{k} \alpha_i y_i(x)$, is a solution of equation (3.29).

The significance of this principle is that the right-hand side of equation (3.29) may be represented as a linear combination of simpler elements, thus reducing the problem of solving the equation to that of solving the somewhat simpler equations (3.33). From the point of view of physics, this means that a compound external action on an object, expressed by the function $f(x)$, may be represented as a superposition of separate elementary actions on this object.

The proof of Theorem 3.3 is based on an identity (valid for k arbitrary n times differentiable functions $u_1, \ldots, u_k$) which follows directly from the properties of derivation:

$$\left(\sum_{i=1}^{k} \alpha_i u_i(x) \right)^{(n)} + a_1(x) \left(\sum_{i=1}^{k} \alpha_i u_i(x) \right)^{(n-1)} + \ldots + a_n(x) \sum_{i=1}^{k} \alpha_i u_i(x)$$

$$= \sum_{i=1}^{k} \alpha_i [u_i^{(n)}(x) + a_1(x) u_i^{(n-1)}(x) + \ldots + a_n(x) u_i(x)]. \quad (3.34)$$

Putting $u_i = y_i(x)$, where $y_i(x)$ is a solution of Eq. (3.33), we obtain for $y(x) = \sum\limits_{i=1}^{k} \alpha_i y_i(x)$

$$y^{(n)}(x) + a_1 y^{(n-1)}(x) + \ldots + a_n(x) y(x) = \sum_{i=1}^{k} \alpha_i f_i(x) = f(x),$$

as required.

Remark. The left hand side of Eq. (3.29) may be viewed as an operator Ly defined on the set of n times differentiable functions y. Then (3.34) means that this operator is linear.

Let us note some important particular cases of Theorem 3.3, stating them as separate assertions.

Theorem 3.4. *Any linear combination of solutions of the homogeneous equation is a solution of the homogeneous equation (this is a particular case of the superposition principle when $f_i = f \equiv 0$).*

Remark. In the language of linear algebra, this may be expressed as follows: the set of solutions of the homogeneous equation is a linear subspace.

Now suppose $k = 2, f_1 = f_2, \alpha_1 = 1, \alpha_2 = -1$ and therefore $f = 0$. Then we have the following:

Theorem 3.5. *The difference between two solutions of the non-homogeneous linear equation satisfies the homogeneous equation.*

In Theorem 3.3 the α_i may also be complex numbers.

Theorem 3.6. *Suppose* $y_1(x), y_2(x)$ *satisfy equation (3.33)* $(i=1,2)$. *Then* $z(x)=y_1(x)+iy_2(x)$ *satisfies the equation*

$$z^{(n)}+a_1(x)z^{(n-1)}+ \ldots +a_n(x)z=f_1+if_2. \qquad (3.35)$$

Conversely: suppose $z(x)=y_1(x)+iy_2(x)$ *satisfies equation (3.35). Then* $y_1(x), y_2(x)$ *satisfy equations (3.33).*

The direct theorem is a particular case of Theorem 3.3 $(\alpha_1=1, \alpha_2=i)$. In order to obtain the converse statement, we must apply identity (3.34) to the left-hand side of (3.35) by setting $u_1=y_1, u_2=y_2, \alpha_1=1, \alpha_2=i$, then set the real part of the expression thus obtained equal to f_1 and its imaginary part equal to f_2, according to the rule for comparing complex numbers.

All the properties stated above are particularly typical of linear equations and considerably simplify their study and solution.

§3. Homogeneous *n*-th Order Linear Equations

Now consider the equation

$$y^{(n)}+a_1(x)y^{(n-1)}+ \ldots +a_n(x)y=0. \qquad (3.36)$$

whose coefficients $a_i(x)$ $(i=1,\ldots,n)$ are continuous on the interval X. As was shown in the previous section, the solution of the initial value problem exists and is unique on X; this will be often used further on.

Definition. We say that the *functions* $u_1(x),\ldots, u_p(x)$ *are linearly dependent on the interval* X if there exist constants $C_1,\ldots, C_p$ not all equal to zero such that we have the identity

$$\sum_{i=1}^{p} C_i u_i(x)=0 \qquad \forall x \in X. \qquad (3.37)$$

In the controverse case, i.e. if (3.37) holds only for $C_1= \ldots =C_p=0$, we shall say that $u_1(x),\ldots, u_p(x)$ are *linearly independent*.

Suppose $y_1(x),\ldots, y_n(x)$ are solutions of equation (3.36).

Definition. The determinant

$$\Delta(x)= \begin{vmatrix} y_1(x) & \ldots y_n(x) \\ y_1'(x) & \ldots y_n'(x) \\ \cdots\cdots\cdots\cdots\cdots \\ y_1^{(n-1)}(x) \ldots y_n^{(n-1)}(x) \end{vmatrix} \qquad (3.38)$$

is known as the *Wronskian*.

(Sometimes the notation $\Delta(y_1(x),\ldots, y_n(x))$ is also convenient.)

Theorem 3.7. *If the solutions $y_1(x), \ldots, y_n(x)$ of equation (3.36) are linearly dependent on X, then $\Delta(x) \equiv 0$ on X.*

Indeed, according to (3.37) we have $\sum_{i=1}^{n} C_i y_i(x) = 0$. Taking derivatives of this identity $n-1$ times, we obtain

$$\sum_{i=1}^{n} C_i y_i(x) = 0, \ldots, \sum_{i=1}^{n} C_i y_i^{(n-1)}(x) = 0. \tag{3.39}$$

For any $x \in X$ these relations may be viewed as a system of linear homogeneous algebraic equations with respect to $C_1, \ldots, C_n$; they have a non-trivial solution in view of the fact that the functions y_i are linearly dependent. Therefore the determinant of the system $\Delta(x)$ vanishes for all $x \in X$, i.e. $\Delta(x) \equiv 0$ on X.

Remark. It is clear from the proof of the theorem that it remains valid not only for solutions of equation (3.36) but for all $n-1$ times differentiable functions.

Theorem 3.8. *If $\Delta(x) = 0$ for at least one $x_0 \in X$, then the solutions $y_1(x), \ldots, y_n(x)$ of equation (3.36) are linearly dependent on X.*

Indeed, take some point $x = x_0$ at which $\Delta(x_0) = 0$ and constitute the system of linear algebraic equations with respect to $C_1, \ldots, C_n$ with determinant $\Delta(x_0)$:

$$\sum_{i=1}^{n} C_i y_i(x_0) = 0, \ldots, \sum_{i=1}^{n} C_i y_i^{(n-1)}(x_0) = 0. \tag{3.40}$$

Since $\Delta(x_0) = 0$, the system possesses a non-trivial solution $C_1, \ldots, C_n$. Consider the linear combination $y(x) = \sum_{i=1}^{n} C_i y_i(x)$. According to Theorem 3.4, $y(x)$ is a solution of Eq. (3.36), while (3.40) means that this solution satisfies the trivial initial conditions $y(x_0) = 0, \ldots, y^{(n-1)}(x_0) = 0$ at the point x_0. Since the trivial solution of Eq. (3.36) $\tilde{y}(x) \equiv 0$ obviously satisfies the same initial conditions, it follows from the uniqueness theorem that $y(x) \equiv \tilde{y}(x) \equiv 0$, i.e. $\sum_{i=1}^{n} C_i y_i(x) \equiv 0$, where, by construction, not all the C_i vanish, but this means that the $y_1(x), \ldots, y_n(x)$ are linearly dependent.

The following alternative is a consequence of the theorems proved above.

Theorem 3.9. *The Wronskian $\Delta(x)$ is either identically zero (and this means that the solutions $y_1(x), \ldots, y_n(x)$ are linearly dependent) or does not vanish at any point of X (and this means that $y_1(x), \ldots, y_n(x)$ are linearly independent).*

The situation may be represented by the following picture

$$\Delta(x) \begin{cases} \text{either} & \Delta(x) \equiv 0 \rightleftarrows y_1(x), \ldots, y_n(x) \text{ linearly dependent} \\ \text{or} & \text{for } \begin{array}{c} \Delta(x) \neq 0 \\ \forall x \in X \end{array} \rightleftarrows y_1(x), \ldots, y_n(x) \text{ linearly independent} \end{cases}$$

Definition. A *fundamental system of solutions* of equation (3.36) is any set of n linearly independent solutions of this equation.

Theorem 3.10. *A fundamental system of solutions exists.*

Indeed, take an arbitrary non-zero determinant Δ^0 with elements a_{ij}. Determine the solutions $y_1(x),\ldots,y_n(x)$ of equation (3.1) by specifying the following initial conditions:

$$y_i(x_0)=a_{1i},\ldots,y_i^{(n-1)}(x_0)=a_{ni}. \tag{3.41}$$

Constitute the Wronskian $\Delta(x)$. By (3.41) if follows that $\Delta(x_0)=\Delta^0 \neq 0$. But then, according to theorem 3.9, the solutions $y_1(x),\ldots,y_n(x)$ are linearly independent.

Remark. Since there exists an infinite set of determinants which are different from zero, for every equation there exists an infinite set of fundamental systems of solutions. Moreover, the linear non-degenerate transformation $\bar{y}_i = \sum_{j=1}^{n} a_{ij} y_j$ sends one fundamental system of solutions into another.

Now let us prove the main theorem of this section.

Theorem 3.11. *Suppose $y_1(x),\ldots,y_n(x)$ is a fundamental system of solutions; then any solution $y(x)$ of equation (3.36) may be represented in the form*

$$y(x)= \sum_{i=1}^{n} C_i y_i(x), \tag{3.42}$$

where the $C_1,\ldots,C_n$ are certain constants.

Proof. Put $y(x_0)=y_1^0,\ldots,y^{(n-1)}(x_0)=y_n^0$. Determine the constants $C_1,\ldots,C_n$ from the linear system of equations with determinant $\Delta(x_0) \neq 0$:

$$\sum_{i=1}^{n} C_i y_i(x_0)=y_1^0,\ldots, \quad \sum_{i=1}^{n} C_i y_i^{(n-1)}(x_0)=y_n^0 \tag{3.43}$$

and construct $\tilde{y}(x)= \sum_{i=1}^{n} C_i y_i(x)$. According to Theorem 3.4, $\tilde{y}(x)$ is the solution of Eq. (3.36), while (3.43) means that this solution satisfies the same initial condition as $y(x)$. Then, by uniqueness (compare with the proof of Theorem 3.8), we have

$$y(x) \equiv \tilde{y}(x)= \sum_{i=1}^{n} C_i y_i(x)$$

which was to be proved.

Remarks. 1. Formula (3.42), where $C_1, \ldots, C_n$ are arbitrary constants, is the general solution of Eq. (3.36) in the same sense as in §1, i.e. (3.42) is a formula which contains all the solutions of Eq. (3.36) and does not contain anything except solutions. Indeed, according to Theorem 3.4, (3.42) for any $C_1, \ldots, C_n$ is the solution of Eq. (3.36), while, according to the theorem that we have just proved, (3.42) contains any solution of Eq. (3.36).

2. In the language of linear algebra, theorems 3.10 and 3.11 mean that the space of solution of the linear homogeneous equation (3.36) possesses a basis consisting of n elements, i.e. this space is n-dimensional.

§4. Non-homogeneous Linear n-th Order Equations

Now consider the equation

$$y^{(n)} + a_1(x)y^{(n-1)} + \ldots + a_n(x)y = f(x), \tag{3.44}$$

where the $a_i(x)$ $(i=1, \ldots, n)$ are continuous in the interval X.

Theorem 3.12. *If $y_1(x), \ldots, y_n(x)$ constitute a fundamental system of solutions of the homogeneous equation (3.36), while $\bar{y}(x)$ is a particular solution of the non-homogeneous equation (3.44), then any solution $y(x)$ of the non-homogeneous equation (3.44) can be represented in the form*

$$y(x) = \bar{y}(x) + \sum_{i=1}^{n} C_i y_i(x), \tag{3.45}$$

where the $C_1, \ldots, C_n$ are certain constants.

Remarks. 1. The theorem is valid for any choice of a particular solution $\bar{y}(x)$.

2. Theorem 3.12 may be stated as follows: the general solution of a non-homogeneous equation is the sum of a particular solution of the non-homogeneous equation and the general solution of the corresponding homogeneous equation.

Proof. Consider the difference $y(x) - \bar{y}(x)$. Then, according to Theorem 3.5, this difference satisfies the homogeneous equation (3.36) and therefore, according to Theorem 3.11, we have

$$y(x) - \bar{y}(x) = \sum_{i=1}^{n} C_i y_i(x).$$

This implies (3.45).

Thus in order to construct the general solution of a non-homogeneous equation, it is necessary to know, besides the fundamental system of solutions of the corresponding homogeneous equation, at least one particular solution of the non-homogeneous equation. Now let us show that if we know the fundamental system of solutions of the homogeneous equation, we can find a particular solution $\bar{y}(x)$ of the non-homogeneous equation in quadratures.

Let us construct particular solution $\bar{y}(x)$ which satisfies the following initial condition

$$\bar{y}(x_0)=0,\ldots,\bar{y}^{(n-1)}(x_0)=0. \tag{3.46}$$

To do this, we shall use the following heuristic consideration. Represent $f(x)$ approximately as the sum of functions (elementary actions) equal to $f(\xi)$ in the interval $(\xi-\varDelta\xi,\xi)$ and equal to zero at the other points. The solution y corresponding to each such elementary action, possessing vanishing derivatives up to the $(n-1)$st order inclusive for $x=x_0$, is identically zero up to $\xi-\varDelta\xi$; however,

$$y^{(n-1)}(\xi)\simeq y^{(n)}(\xi-\varDelta\xi)\varDelta\xi=[y^{(n)}(\xi-\varDelta\xi)\varDelta\xi$$
$$+a_1 y^{(n-1)}(\xi-\varDelta\xi)\varDelta\xi+\ldots+a_n y(\xi-\varDelta\xi)\varDelta\xi]=f(\xi-\varDelta\xi)\varDelta\xi,$$

i.e. $y^{(n-1)}(\xi)$ no longer equals zero but equals $f(\xi)\varDelta\xi$ and therefore the solution from then on will not be zero. According to the superposition principle, it is sufficient to construct a solution of the homogeneous equation (since outside $(\xi-\varDelta\xi,\xi)$ the right-hand side vanishes) which assumes a zero value at the point ξ together with its derivatives up to the $(n-2)$ order inclusive and has an $(n-1)$-st order derivative equal to one (let us denote the solution by $\mathcal{K}(x,\xi)$, thus indicating its dependence on the initial point, and call it the *impulse function*) and then multiply it by $f(\xi)\varDelta\xi$. Thus $\mathcal{K}(x,\xi)$ is constructed as the solution of the homogeneous equation satisfying the conditions

$$\mathcal{K}(\xi,\xi)=0,\ldots,\mathcal{K}_x^{(n-2)}(\xi,\xi)=0,\ \mathcal{K}_x^{(n-1)}(\xi,\xi)=1, \tag{3.47}$$

while the solution corresponding to the elementary action will be of the form

$$\mathcal{K}(x,\xi)f(\xi)\varDelta\xi.$$

Bringing together all the elementary actions, using the superposition principle again and passing from the sum to the integral, we shall obtain the solution $\bar{y}(x)$ which satisfies condition (3.46):

$$\bar{y}(x)=\int_{x_0}^{x}\mathcal{K}(x,\xi)f(\xi)d\xi. \tag{3.48}$$

Formula (3.48) was obtained on the basis of heuristic considerations, but it is not difficult to verify directly that (3.48) is a particular solution of equation (3.44). This verification will be the proof of the next theorem.

Theorem 3.13. *The expression (3.48), where the function $\mathcal{K}(x, \xi)$, called the impulse function, satisfies the homogeneous equation (3.36) and the initial condition (3.47) is a particular solution of the non-homogeneous equation (3.44) satisfying the zero initial conditions (3.46).*

Indeed, let us find $\bar{y}', \ldots, \bar{y}^{(n)}$ from (3.48). Firstly, since ξ is a parameter belonging to the same set as x, it follows that (3.47) is equivalent to the following relations

$$\mathcal{K}(x, x) = 0, \ldots, \mathcal{K}_x^{(n-2)}(x, x) = 0, \quad \mathcal{K}_x^{(n-1)}(x, x) = 1.$$

Taking derivatives in (3.48), we get

$$\bar{y}'(x) = \int_{x_0}^{x} \mathcal{K}_x'(x, \xi) f(\xi) d\xi + \mathcal{K}(x, x) f(x) = \int_{x_0}^{x} \mathcal{K}_x'(x, \xi) f(\xi) d\xi,$$

$$\cdots\cdots\cdots\cdots\cdots\cdots\cdots\cdots\cdots\cdots\cdots\cdots\cdots\cdots\cdots\cdots\cdots\cdots$$

$$\bar{y}^{(n-1)}(x) = \int_{x_0}^{x} \mathcal{K}_x^{(n-1)}(x, \xi) f(\xi) d\xi,$$

$$\bar{y}^{(n)}(x) = \int_{x_0}^{x} \mathcal{K}_x^{(n)}(x, \xi) f(\xi) d\xi + \mathcal{K}_x^{(n-1)}(x, x) f(x)$$

$$= \int_{x_0}^{x} \mathcal{K}_x^{(n)}(x, \xi) f(\xi) d\xi + f(x).$$

It is possible to take derivatives under the integral sign because of the theorem on the continuous dependence of the solution of systems of differential equations on x and on the initial value of the variable x, i.e. in the given case on ξ (see §5 in Chapter 2).

Substituting the previous relation into (3.44), we obtain

$$\bar{y}^{(n)} + a_1(x) \bar{y}^{(n-1)} + \ldots + a_n \bar{y} = \int_{x_0}^{x} [\mathcal{K}_x^{(n)}(x, \xi) + a_1(x) \mathcal{K}_x^{(n-1)}(x, \xi) + \ldots$$

$$+ a_n(x) \mathcal{K}(x, \xi)] f(\xi) d\xi + f(x) = f(x),$$

since the square bracket under the integral sign vanishes in view of the definition of $\mathcal{K}(x, \xi)$. Thus $\bar{y}(x)$ is indeed the solution of Eq. (3.44) and, moreover, obviously satisfies conditions (3.46).

Remark. In particular, for first order equations, formula (3.48) coincides with formula (2.29) if we put $y_0 = 0$. In (2.29) the impulse function is the multiplier $e^{-\int_{\xi}^{x} p(\xi) d\xi}$ which, according to (2.26), satisfies the homogeneous equation and becomes equal to 1 when $x = \xi$.

§5. Linear n-th Order Equations with Constant Coefficients

If we know a fundamental system of solutions, it is possible to find any solution of the homogeneous equation and, by applying quadratures, to find the solution of the non-homogeneous equation. The existence of fundamental systems of solutions was proved (see Theorem 3.10); however, the question of their effective construction remains open.

Suppose in (3.36) the a_i are constant

$$y^{(n)} + a_1 y^{(n-1)} + \ldots + a_n y = 0. \tag{3.49}$$

This class of equations is remarkable since for it a fundamental system of solutions may be found by using algebraic operations only, in fact, the problem reduces to the solution of an algebraic equation of the n-th degree.

To equation (3.49) let us assign a polynominal in λ, called the characteristic polynomial of equation (3.49):

$$M(\lambda) = \lambda^n + a_1 \lambda^{n-1} + \ldots + a_n.$$

Lemma 3.1. *We have the identity*

$$\frac{d^n}{dx^n} [e^{\lambda x} f(x)] + a_1 \frac{d^{n-1}}{dx^{n-1}} [e^{\lambda x} f(x)] + \ldots + a_n e^{\lambda x} f(x)$$

$$= e^{\lambda x} \left\{ M(\lambda) f(x) + M'(\lambda) f'(x) \right.$$

$$\left. + \frac{M''(\lambda) f''(x)}{2!} + \ldots + \frac{M^{(n)}(\lambda) f^{(n)}(x)}{n!} \right\}. \tag{3.50}$$

This identity is proved by direct computation, using the Leibnitz formula for the derivative of a product. We have

$$e^{\lambda x} f = e^{\lambda x} f,$$

$$\frac{d}{dx} (e^{\lambda x} \cdot f) = e^{\lambda x} (\lambda f + f') = e^{\lambda x} (\lambda f + \lambda' f'),$$

$$\frac{d^2}{dx^2} (e^{\lambda x} \cdot f) = e^{\lambda x} (\lambda^2 f + 2\lambda f' + f'') = e^{\lambda x} \left(\lambda^2 f + \frac{(\lambda^2)' f'}{1!} + \frac{(\lambda^2)'' f''}{2!} \right),$$

$$\frac{d^n}{dx^n} (e^{\lambda x} \cdot f) = e^{\lambda x} \left(\lambda^n f + \ldots + \frac{n(n-1)\ldots[n-(k-1)]}{k!} \lambda^{n-k} f^{(k)} + \ldots + f^{(n)} \right)$$

$$= e^{\lambda x} \left(\lambda^n f + (\lambda^n)' f' + \ldots + \frac{(\lambda^n)^{(k)} f^{(k)}}{k!} + \ldots + \frac{(\lambda^n)^{(n)} f^{(n)}}{n!} \right).$$

Adding these relations multiplied by the corresponding a_i, we obtain (3.50).

Remarks. 1. If $f(x) = x^p$, then (3.50) acquires the form

$$\frac{d^n}{dx^n}(e^{\lambda x}x^p) + a_1\frac{d^{n-1}}{dx^{n-1}}(e^{\lambda x}x^p) + \ldots + a_n e^{\lambda x}x^p = e^{\lambda x}\left\{M(\lambda)x^p + pM'(\lambda)x^{p-1}\right.$$

$$\left. + \ldots + \frac{p\ldots(p-k+1)}{k!}M^{(k)}(\lambda)x^{p-k} + \ldots + M^{(p)}(\lambda)\right\}. \qquad (3.51)$$

In particular, when $p = 0$, we have

$$\frac{d^k}{dx^k}e^{\lambda x} + a_1\frac{d^{k-1}}{dx^{k-1}}e^{\lambda x} + \ldots + a_n e^{\lambda x} = e^{\lambda x}M(\lambda). \qquad (3.52)$$

2. The identities (3.50)–(3.52) may be written in more compact form if we denote the differential operator by $D: \dfrac{dy}{dx} = Dy$. If we use the usual rules for adding and multiplying operators, then the left-hand side of equation (3.49) may be written in the form

$$(D^n + a_1 D^{n-1} + \ldots + a_n)y = M(D)y.$$

The operator $M(D)$ is said to be a polynomial operator. It has the same structure as the characteristic polynomial $M(\lambda)$.

Introducing $M(D)$, we can rewrite the identities (3.50)–(3.52) in the form

$$M(D)e^{\lambda x}f(x) = e^{\lambda x}\left\{M(\lambda)f(x) + M'(\lambda)f'(x) + \ldots + \frac{M^{(n)}(\lambda)f^{(n)}(x)}{n!}\right\},$$

$$\qquad (3.53)$$

$$M(D)e^{\lambda x}x^p = e^{\lambda x}\left\{M(\lambda)x^p + \ldots + \frac{p\ldots(p-k+1)}{k!}M^{(k)}(\lambda)x^{p-k} + \ldots + M^{(p)}(\lambda)\right\},$$

$$\qquad (3.54)$$

$$M(D)e^{\lambda x} = e^{\lambda x}M(\lambda). \qquad (3.55)$$

Also note the following property of polynomial operators which will be needed later. Consider, together with $M(D)$, another polynomial operator $N(D) = D^s + b_1 D^{s-1} + \ldots + b_s$ (the b_s are constants). Using the rule for adding and multiplying operators, it is easy to verify that polynomial operators are multiplied according to the rule of multiplication of ordinary polynomials

$$M(D)N(D) = N(D)M(D) = D^{n+s} + (a_1 + b_1)D^{n+s-1} + \ldots + a_n b_s.$$

Putting $M(\lambda)$ equal to zero, we obtain an algebraic equation of the n-th degree with respect to λ, the so-called characteristic equation

$$\lambda^n + a_1 \lambda^{n-1} + \ldots + a_n = 0. \tag{3.56}$$

Assume that this equation has the roots $\lambda_1, \ldots, \lambda_l$ of multiplicities $m_1, \ldots, m_l$ $(m_1 + \ldots + m_l = n)$.

Theorem 3.14. *1. To the root λ_k of the characteristic equation (3.56) of multiplicity m_k correspond m_k particular solutions of the form*

$$e^{\lambda_k x}, \ xe^{\lambda_k x}, \ldots, x^{m_k-1}e^{\lambda_k x}. \tag{3.57}$$

2. The solutions (3.57), where $k = 1, \ldots, l$, constitute a fundamental system of solutions of equation (3.49).

Proof. 1. Let us use (3.51) or (3.54). If λ_k is a root of the characteristic equation of multiplicity m_k, then

$$M(\lambda_k) = M'(\lambda_k) = \ldots = M^{(m_k-1)}(\lambda_k) = 0.$$

(This follows from the fact that $M(\lambda)$ may be written in the form $M(\lambda) = (\lambda - \lambda_k)^{m_k} N(\lambda)$, where $N(\lambda)$ is a polynomial of degree $n - m_k$.)

Therefore, the right-hand side of (3.51) vanishes for $p = 0, 1, \ldots, m_k - 1$ and this means that $x^p e^{\lambda_k x}$ $(p = 0, 1, \ldots, m_k - 1)$ satisfies Eq. (3.49), as required.

2. Assume the converse, i.e. assume that the solutions of (3.57) $(k = 1, \ldots, l)$ are linearly dependent. This means that we have the identity

$$R_1(x)e^{\lambda_1 x} + \ldots + R_l(x)e^{\lambda_l x} = 0. \tag{3.58}$$

where by $R_j(x)$ we have denoted polynomials of degree $m_j - 1$ which are not all identically zero. Assume that R_1 is not identically zero (this may be achieved by an appropriate renumbering of the λ) and that the highest non-zero term in R_1 is of degree p_1 $(p_1 \leq m_1 - 1)$, i.e.

$$R_1(x) = C_{10} + C_{11}x + \ldots + C_{1p_1}x^{p_1},$$

where $C_{1p_1} \neq 0$.

Multiply (3.58) by $e^{-\lambda_1 x}$. Then we obtain

$$R_1(x)e^{(\lambda_1 - \lambda_1)x} + \ldots + R_{l-1}(x)e^{(\lambda_{l-1} - \lambda_l)x} + R_l(x) = 0. \tag{3.59}$$

Take the derivative of this identity $p_l + 1$ times, where $p_l \leq m_l - 1$ is the degree of the polynomial $R_l(x)$. Note first that in the expression $A(x)e^{\alpha x}$, where $\alpha = \text{const}$, $A(x)$ is a polynomial and for an arbitrary k we have the identity

$$\frac{d^k}{dx^k} A(x)e^{\alpha x} = B(x)e^{\alpha x},$$

where $B(x)$ is a polynomial of the same degree as $A(x)$, and its highest order coefficient is equal to the highest order coefficient of $A(x)$ multiplied by α^k. This identity can easily be obtained either from (3.53), by putting $M(D) = D^k$, $\lambda = \alpha$, $f(x) = A(x)$ or simply from the Leibnitz formula. Thus, taking derivatives of (3.59), we obtain

$$Q_1(x)e^{(\lambda_1 - \lambda_l)x} + \ldots + Q_{l-1}(x)e^{(\lambda_{l-1} - \lambda_l)x} = 0,$$

or

$$Q_1(x)e^{\lambda_1 x} + \ldots + Q_{l-1}(x)e^{\lambda_{l-1} x} = 0, \qquad (3.60)$$

where $Q_1(x), \ldots, Q_{l-1}(x)$ are polynomials of the same degree as $R_1, \ldots, R_{l-1}$, and the coefficient of the highest order term of $Q_1(x)$ is $C_{1 p_1}(\lambda_1 - \lambda_l)^{p_l + 1}$. If we now apply to (3.60) the same operation we just applied to (3.58) and continue this process, we finally obtain an identity of the form

$$S_1(x)e^{\lambda_1 x} = 0, \quad \text{or} \quad S_1(x) = 0, \qquad (3.61)$$

where the coefficient of the highest order term of $S_1(x)$ is $C_{1 p_1}(\lambda_1 - \lambda_l)^{p_l + 1} \cdot \ldots \cdot (\lambda_1 - \lambda_2)^{p_2 + 1}$ and, by (3.61), it must vanish; this contradicts the fact that $C_{1 p_1} \neq 0$, $\lambda_1 - \lambda_l \neq 0$, $\ldots$, $\lambda_1 - \lambda_2 \neq 0$. The contradiction just obtained implies that the solutions of (3.57) are linearly independent, i.e. constitute a fundamental system of solutions; statement 2 is therefore proved.

In view of Theorem 3.14, the general solution of equation (3.49) is a linear combination of the solutions of (3.57) ($k = 1, \ldots, l$). But in the case of complex λ_k this representation is not always convenient. However, instead of the fundamental system of solutions (3.57), it is possible to use another fundamental system of solutions consisting of real-valued functions.

Suppose $\lambda_k = p_k + iq_k$. Then, to the two complex solutions of the form $x^r e^{\lambda_k x}$, $x^r e^{\lambda_k^* x}$ (where λ_k^* denotes a root of the characteristic equations which is complex conjugate to λ_k; this root exists in view of the fact that the coefficients of the characteristic equation are real) correspond, in view of Theorem 3.6, the two real solutions

$$\text{Re } (x^r e^{\lambda_k x}) = x^r e^{p_k x} \cos q_k x \quad \text{and} \quad \text{Im } (x^r e^{\lambda_k x}) = x^r e^{p_k x} \sin q_k x.$$

Thus, instead of complex solutions, we may construct the same number of real solutions; they constitute another fundamental system of solutions – they are linearly independent, since the matrix transforming the pair of complex conjugate solutions into their real and imaginary parts is of the form $\begin{pmatrix} 1 & i \\ 1 & -i \end{pmatrix}$ and has a non-zero determinant (equal to $-2i$).

Taking the linear combination of the real solutions just obtained, we get the representation (3.25), which is therefore established.

Now consider the non-homogeneous equation

$$y^{(n)} + a_1 y^{(n-1)} + \ldots + a_n y = f(x). \tag{3.62}$$

Knowing the fundamental system of solutions (3.57), we can construct its particular solution according to Theorem 3.13. However in practice, this requires certain complicated calculations; it is therefore of interest to study the class of functions $f(x)$ for which a particular solution may be constructed without using formula (3.48), by means of purely algebraic operations.

Theorem 3.15. *Suppose $f(x) = S(x)e^{\lambda x}$, where $\lambda = const$, $S(x)$ is a polynomial of degree s. Suppose λ does not coincide with any other root λ_k of the characteristic equation (3.56) (the so-called non-resonance case). Then there exists a particular solution of equation (3.62) of the form*

$$\bar{y}(x) = P(x)e^{\lambda x}, \tag{3.63}$$

where $P(x)$ is a polynomial of the same degree as $S(x)$. If λ coincides with a root of the characteristic equation λ_k of multiplicity m_k (the so-called resonance case), then there exists a particular solution of equation (3.62) of the form

$$\bar{y}(x) = T(x)x^{m_k}e^{\lambda x}, \tag{3.64}$$

where $T(x)$ is a polynomial of the same degree as $S(x)$.

On the basis of this theorem, a particular solution may be looked for in the form indicated above, where the polynomials $P(x)$ or $T(x)$ are written out with unknown coefficients. Substituting them into equation (3.62), dividing by $e^{\lambda x}$ and setting the terms in identical powers of x equal to each other, we obtain a system of non-homogeneous algebraic equations with respect to the unknown coefficients of the polynomials $P(x)$ and $T(x)$. The system can be solved, since the existence of solutions of this form is guaranteed by Theorem 3.15.

The proof of Theorem 3.15 will be carried out for the resonance case (3.66), since (3.63) is obtained from (3.64) by putting $m_k = 0$. Let us substitute (3.64) into (3.62):

$$M(D)T(x)x^{m_k}e^{\lambda x} = e^{\lambda x}S(x); \tag{3.65}$$

we then see that this allows us to successively determine the coefficients of the polynomial $T(x)$, beginning with the coefficient of the highest order term x^s. Let us display the highest order terms in the polynomials $T(x)$ and $S(x)$:

$$S(x) = a_0 x^s + S_1(x), \qquad T(x) = b_0 x^s + T_1(x).$$

We then have

$$M(D)b_0 x^{s+m_k}e^{\lambda x} + M(D)x^{m_k}T_1(x)e^{\lambda x} = e^{\lambda x}a_0 x^s + e^{\lambda x}S_1(x).$$

Writing out the first summand from the left, using formula (3.54) and taking into consideration the fact that $M(\lambda) = M'(\lambda) = \ldots = M^{(m_k-1)}(\lambda) = 0$, while $M^{(m_k)}(\lambda) \neq 0$, we get

$$b_0 e^{\lambda x} \left\{ \frac{M^{(m_k)}(\lambda)(s+m_k)\ldots(s+1)}{m_k!} x^s + \frac{M^{(m_k+1)}(\lambda)(s+m_k)\ldots s}{(m_k+1)!} x^{s-1} + \ldots \right\}$$

$$+ M(D) x^{m_k} T_1(x) e^{\lambda x} = e^{\lambda x} a_0 x^s + e^{\lambda x} S_1(x). \tag{3.66}$$

Note that the first summand in the figure brackets is of degree s, while the others have lower degrees. Setting the higher order terms equal to each other and dividing by $e^{\lambda x} x^s$, we get

$$b_0 M^{(m_k)}(\lambda) \frac{(s+m_k)\ldots(s+1)}{m_k!} = a_0.$$

This enables us to determine b_0 in terms of a_0, since $M^{(m_k)}(\lambda) \neq 0$. After this (3.66) may be rewritten in the form

$$M(D) x^{m_k} T_1(x) e^{\lambda x} = e^{\lambda x} \tilde{S}_1(x), \tag{3.67}$$

where $\tilde{S}_1(x)$ is a polynomial of degree no greater than $s-1$ obtained by transferring all the terms of the expression $b_0 e^{\lambda x} \{.\}$ (except the first one) which are known to the right-hand side.

(3.67) is an equation similar to (3.65), but the degrees of the polynomials $T_1(x)$ and $\tilde{S}_1(x)$ are one less than those of $T(x)$ and $S(x)$. From (3.67), similarly to the previous case, we obtain the highest order coefficient of the polynomial $T_1(x) = a_1 x^{s-1} + T_2(x)$; we have determined the two highest order coefficients of the polynomial $T(x)$. Continuing this process, we can determine all the terms of $T(x)$ one after the other.

The method for finding a particular solution based on the theorem proved above, will be referred to as the indetermined coefficient method.

Thus, for an equation with constant coefficients, a fundamental system of solutions and, in the case of a right-hand side of the form $e^{\lambda x} S(x)$ a particular solution of the non-homogeneous equation as well, may be constructed in effective form by means of algebraic operations. In conclusion, let us indicate a special class of equations with variable coefficients for which a fundamental system of solutions may also be constructed effectively. This is the so-called Euler equation

$$x^n a_0 y^{(n)} + x^{n-1} a_1 y^{(n-1)} + \ldots + a_n y = 0, \quad a_i = \text{const}. \tag{3.68}$$

By direct calculation it is easy to verify that the change of independent variable $x = e^s$ transforms equation (3.68) into an equation with constant coefficients, thus yielding an effective construction of a fundamental system of solutions.

In order to find a particular solution of the non-homogeneous Euler equation in the case when the right-hand side is of the form $x^{\lambda}S(\log x)$, one may apply the method of indetermined coefficients.

§6. Systems of Linear Equations. General Theory

Let us consider a system of linear differential equations

$$y_i' = \sum_{k=1}^{n} a_{ik}(x)y_k + f_i(x) \qquad (i=1,\ldots,n). \tag{3.69}$$

The system (3.69) is said to be homogeneous if $f_i(x)=0$ $(i=1,\ldots,n)$ and non-homogeneous in the converse case. We shall assume that the functions $a_{ik}(x)$ and $f_i(x)$ are continuous on the interval X. As was proved above (see §2), under these conditions there exists a unique solution of the system (3.69) on X satisfying the initial condition

$$y_i(x_0) = y_i^0 \qquad (i=1,\ldots,n). \tag{3.70}$$

For systems of equations, we have theorems similar to those which were proved for one equation of the n-th order.

1. Matrix notation. In order to simplify the formulas used in our exposition, it shall be convenient to use matrix notation. Let us recall the main facts from the calculus of matrices which will be necessary to us.

$1°$. A matrix of dimension $n \times m$ (or an $(n \times m)$-matrix) is a table of numbers a_{ik} of the form

$$\begin{pmatrix} a_{11} & a_{12} & \cdots & a_{1m} \\ \cdots & \cdots & \cdots \\ a_{n1} & a_{n2} & \cdots & a_{nm} \end{pmatrix}.$$

The numbers a_{ik} are called elements of the matrix.

In the present section, we shall use square matrices (in other words $(n \times n)$-matrices) and so-called column matrices (or $(n \times 1)$-matrices)

$$\begin{pmatrix} a_{11} \\ \vdots \\ a_{n1} \end{pmatrix} \quad \text{or simply} \quad \begin{pmatrix} a_1 \\ \vdots \\ a_n \end{pmatrix}.$$

We shall denote matrices by a, b, c, etc. and their elements respectively by a_{ij}, b_{ij}, c_{ij}, etc.

$2°$. Two matrices a and b are considered equal if $a_{ij}=b_{ij}$. The matrix a is considered equal to zero if $a_{ij}=0$.

$3°$. The operations of addition and multiplication by numbers are defined for $(n \times m)$-matrices.

The sum of the matrices a and b is the matrix c (denoted by $c=a+b$) such that $c_{ij}=a_{ij}+b_{ij}$.

The product of the matrix a by the number α is the matrix c (denoted by $c=\alpha a$) such that $c_{ij}=\alpha a_{ij}$.

$4°$. If a is an $(n \times m)$-matrix, while b is an $(m \times p)$-matrix, then the product of matrices a and b is by definition the matrix c of dimension $n \times p$ (denoted by $c=ab$) such that

$$c_{ij}= \sum_{l=1}^{m} a_{il}b_{lj}.$$

Multiplication of matrices possesses the properties of associativity and distributivity.

For square matrices of the same dimension, both the product ab and the product ba are defined, but in general the multiplication of matrices is not commutative, i.e. $ab \neq ba$.

$5°$. The matrix inverse to the $(n \times n)$-matrix a is the matrix c (denoted by $c=a^{-1}$) such that $c_{ij}=\dfrac{1}{\Delta} A_{ji}$, where $\Delta=\mathrm{Det}\, a$ while A_{ij} is the cofactor of the element a_{ij}. We have the following relations: $aa^{-1}=a^{-1}a=E$, where

$$E= \begin{pmatrix} 1 \dots 0 \\ \dots \dots \\ 0 \dots 1 \end{pmatrix}$$

is the so-called unit matrix, whose only non-zero elements are units standing along the main diagonal.

We shall also consider matrices whose elements are functions of x. For such matrices, besides the operations listed above, the operations of the calculus – differentiation and integration – are also defined.

$6°$. The derivative $a'(x)$ of the matrix $a(x)$ with elements $a_{ij}(x)$ is the matrix with elements $a'_{ij}(x)$ (by definition). The rules for taking the derivative of a sum or a product remain valid for matrices, except that in taking the derivative of the product of matrices it is necessary to preserve the order of the factors:

$$(ab)' =a'b+ab'.$$

$7°$. $\int\limits_{x_0}^{x} a(x)dx$ is determined as the matrix with elements

$$\int\limits_{x_0}^{x} a_{ij}(x)dx.$$

2. General properties of systems of linear equations. Let us consider the system (3.69). Denote by y, f and y^0 the columns

$$y = \begin{pmatrix} y_1 \\ \cdot \\ \cdot \\ y_n \end{pmatrix}, \quad f = \begin{pmatrix} f_1 \\ \cdot \\ \cdot \\ f_n \end{pmatrix}, \quad y^0 = \begin{pmatrix} y_1^0 \\ \cdot \\ \cdot \\ y_n^0 \end{pmatrix},$$

and denote the $(n \times n)$-matrix with elements $a_{ij}(x)$ by $A(x)$:

$$A(x) = \begin{pmatrix} a_{11}(x) \ldots a_{1n}(x) \\ \cdot \quad \cdot \quad \cdot \quad \cdot \quad \cdot \\ a_{n1}(x) \ldots a_{nn}(x) \end{pmatrix}.$$

Then the system (3.69) may be written in the form of one equation

$$y' = A(x)y + f(x) \tag{3.71}$$

just as the initial conditions

$$y(x_0) = y^0. \tag{3.72}$$

Using the rule of multiplication 4°, the rule of addition and equality of matrices 3°, 2°, it is easy to verify that (3.71) and (3.72) are the same as (3.69) and (3.70).

In view of the properties of multiplication and differentiation of matrices, we have the following identity for the differentiation of columns (compare with (3.34)), in which the α_i are constants,

$$\left(\sum_{i=1}^{k} \alpha_i u_{(i)} \right)' - A \left(\sum_{i=1}^{k} \alpha_i u_{(i)} \right) = \sum_{i=1}^{k} \alpha_i (u'_{(i)} - A u_{(i)}), \tag{3.73}$$

expressing the linearity property of the operator $y' - Ay \equiv L[y]$ on the set of differentiable columns.

Here and later we shall use an index placed in parentheses to number columns, leaving indices without parentheses to denote elements (components).

A direct corollary of identity (3.73) is the superposition principle.

Theorem 3.16. *Suppose that $f(x)$ in equation (3.71) is a linear combination of the $f_{(i)}(x)$, i.e. $f(x) = \sum_{i=1}^{k} \alpha_i f_{(i)}(x)$, where the α_i are constants and suppose $y_{(i)}(x)$ is a solution of the equation $y'_{(i)} = A(x) y_{(i)} + f_{(i)}$. Then the linear combination of the $y_{(i)}(x)$ with the same coefficients α_i, i.e. the column $y(x) = \sum_{i=1}^{k} \alpha_i y_{(i)}(x)$ is the solution of equation (3.71).*

Theorems similar to Theorems 3.4–3.6 are also valid.

3. Homogeneous equations. Consider in more detail the homogeneous equation

$$y' = A(x)\, y. \tag{3.74}$$

Suppose we have n columns

$$y_{(i)} = \begin{pmatrix} y_{(i)1} \\ \cdot \\ \cdot \\ \cdot \\ y_{(i)n} \end{pmatrix}$$

From these columns let us form the matrix $W(x)$:

$$W(x) = \begin{pmatrix} y_{(1)1}(x) \dots y_{(n)1}(x) \\ \dots \qquad \dots \\ y_{(1)n}(x) \dots y_{(n)n}(x) \end{pmatrix}. \tag{3.75}$$

For equation (3.74), whose right-hand and left-hand sides are columns, define a similar equation

$$W' = A(x)\, W, \tag{3.76}$$

whose right- and left-hand sides are $(n \times n)$-matrices and in which the unknown is the matrix $W(x)$.

Theorem 3. 17. *Suppose $y_{(1)}, \dots, y_{(n)}$ are n solutions of equation (3.74). Then the $(n \times n)$-matrix $W(x)$ constructed from these solutions by means of formula (3.75) is a solution of the matrix equation (3.76). Conversely: if $W(x)$ is a solution of equation (3.76), then each column of the matrix $W(x)$ is a solution of equation (3.74).*

In order to verify this, it is sufficient to write out (3.76) and (3.74) element by element. Indeed, (3.76) means

$$W'_{ij} = \sum_{k=1}^{n} a_{ik} W_{kj}, \tag{3.77}$$

or

$$y'_{(j)i} = \sum_{k=1}^{n} a_{ik} y_{(j)k}, \tag{3.78}$$

while (3.74) means

$$y'_i = \sum_{k=1}^{n} a_{ik} y_k. \tag{3.79}$$

Therefore, if $y_{(j)}$ is a solution of (3.74), then each $y_{(j)}$ satisfies (3.79), i.e. we have (3.78) or (3.77), which is the same thing, and therefore we also have (3.76); conversely, if we have (3.76), then we also have (3.78), which, taken together with (3.79), means that $y_{(j)}$ $(j=1,\ldots,n)$ is a solution of equation (3.74).

We can also note the following facts, which may be verified just as easily.

Theorem 3.18. *Suppose $W(x)$ is a solution of equation (3.76); then the expression WB is a solution of equation (3.74), if B is an arbitrary constant column, and the solution of equation (3.76), if B is an arbitrary constant $(n \times n)$-matrix.*

Definition. We shall say that the columns $u_{(1)},\ldots,u_{(p)}$ are linearly dependent on the interval X, if there exist constants $C_1,\ldots,C_p$, not all equal to zero, such that we have the identity

$$\sum_{i=1}^{p} C_i u_{(i)}(x) \equiv 0, \qquad x \in X. \tag{3.80}$$

If (3.80) holds only in the case when $C_1 = \ldots = C_p = 0$, then we shall say that $u_{(1)},\ldots,u_{(p)}$ are linearly independent.

Consider n differentiable columns $y_{(1)},\ldots,y_{(n)}$. Write out equality (3.80) for them:

$$\sum_{i=1}^{n} C_i y_{(i)}(x) \equiv 0. \tag{3.81}$$

Introduce the constant column

$$C = \begin{pmatrix} C_1 \\ \vdots \\ C_n \end{pmatrix}.$$

Using this column and the matrix $W(x)$ constructed from the $y_{(i)}$ according to the rule (3.75), we can rewrite (3.81) in the form

$$WC = 0. \tag{3.82}$$

If we now take into consideration rule $2°$ of matrix calculus, we see that $C=0$ if all the C_i $(i=1,\ldots,n)$ vanish, so that the definition of linear dependence and independence of $y_{(1)},\ldots,y_{(n)}$ may be stated in the following way.

Definition. We shall say that the columns $y_{(1)},\ldots,y_{(n)}$ are *linearly dependent* on the interval X if there exists a constant column $C \neq 0$ such that we have (3.82) identically on X.

Conversely, i.e. when (3.82) is valid only in the case $C=0$, we shall say that the $y_{(1)},\ldots,y_{(n)}$ are linearly independent.

Definition. Let us call $\varDelta(x) = \operatorname{Det} W(x)$ the *Wronskian* for $y_{(1)}, \ldots, y_{(n)}$.

Now we can state and prove theorems similar to Theorems 3.7–3.9 from the theory of n-th order equations. All the proofs can now be written in very compact form if we use the matrix notation introduced above, which is very convenient and easy to get used to.

Theorem 3.19. *If the solutions $y_{(1)}, \ldots, y_{(n)}$ of equation (3.74) are linearly dependent on X, then $\varDelta(x) \equiv 0$ on X.*

Indeed, we have $WC = 0$, $C \neq 0$. This notation is a brief expression of the fact that for every x the values $C_1, \ldots, C_n$ satisfy a system of linear algebraic equations with determinant $\varDelta(x)$ and, since the solution is non-trivial, it follows that $\varDelta(x) = 0$ for any $x \in X$, i.e. $\varDelta(x) \equiv 0$.

Theorem 3.20. *If $\varDelta(x) = 0$ for at least one $x_0 \in X$, then the solutions $y_{(1)}, \ldots, y_{(n)}$ of equation (3.74) are linearly dependent on X.*

Let us briefly write out the proof of this theorem, no longer giving supplementary explanations as we did in the previous proof. Choose $x_0 \in X$ and suppose $\varDelta(x_0) = 0$. Constitute equation $W(x_0) C = 0$ with respect to C. Since $\varDelta(x_0) = 0$ we have a non-zero solution C. Put $y(x) = W(x) C$. According to Theorem 3.18, this is a solution of Eq. (3.74) and we have $y(x_0) = W(x_0) C = 0$ and according to the uniqueness theorem, $y(x) \equiv 0$, so that finally $W(x) C \equiv 0$, this means that $y_{(1)}, \ldots, y_{(n)}$ are linearly dependent.

Theorem 3.21 *(the alternative). The Wronskian is either identically zero (and this means that the solutions $y_{(1)}, \ldots, y_{(n)}$ are linearly dependent) or does not vanish at any point of X (and this means that the solutions $y_{(1)}, \ldots, y_{(n)}$ are linearly independent).*

Definition. A *fundamental system of solutions* of equation (3.74) is by definition a set of n linearly independent solutions $y_{(1)}, \ldots, y_{(n)}$ of Eq. (3.74), while the matrix $W(x)$ which corresponds to them according to formula (3.75) will be called the *fundamental matrix*.

On the basis of Theorem 3.20, we can give another (equivalent) definition of the fundamental matrix.

Definition. The solution $W(x)$ of equation (3.76) for which $\varDelta(x)$ does not vanish everywhere on X is called the fundamental matrix.

Theorem 3.22. *A fundamental matrix exists.*

According to Theorem 3.21, it is sufficient to take an arbitrary matrix $a = \text{const}$ with non-zero determinant and choose for W the initial condition $W(x_0) = a$.

Theorem 3.23. *If $W(x)$ is a fundamental matrix, then any solution $y(x)$ of equation (3.74) may be represented in the form*

$$y(x) = W(x)C, \tag{3.83}$$

where C is a certain constant column.

Proof. Suppose $y(x_0) = y^0$. Determine C from the equation $W(x_0)C = y^0$, which possesses a solution since $\Delta(x_0) \neq 0$. Construct $\tilde{y}(x) = W(x)C$. Since $\tilde{y}(x_0) = W(x_0)C = y^0$, it follows by the uniqueness theorem that $y(x) \equiv \tilde{y}(x) = W(x)C$ as required.

Remark. In the language of linear algebra, Theorems 3.22 and 3.23 mean that the space of solutions of equation (3.74) is n-dimensional.

Construct the solution of equation (3.74) satisfying condition (3.72) by expressing C in terms of y^0 with the help of $W(x)$. We have

$$y(x_0) = W(x_0)C = y^0,$$

hence $C = W^{-1}(x_0)y^0$ and therefore

$$y(x) = W(x)W^{-1}(x_0)y^0.$$

The matrix $\mathcal{K}(x, x_0) = W(x)W^{-1}(x_0)$, which is a function of two variables x and x_0, is said to be the *impulse matrix* (compare with §4) or the *matrizant*. According to Theorem 3.18, $\mathcal{K}(x, x_0)$ as a function of x satisfies equation (3.76). Moreover, we obviously have

$$\mathcal{K}(x_0, x_0) = E.$$

Therefore, we have the following

Theorem 3.24. *The solution of the problem (3.74), (3.72) is of the form*

$$y(x) = \mathcal{K}(x, x_0)y^0, \tag{3.84}$$

where the matrix $\mathcal{K}(x, x_0)$, called the impulse matrix or matrizant, satisfies (with respect to its variable x) the matrix equation (3.76) and the condition $\mathcal{K}(x_0, x_0) = E$.

4. Non-homogeneous equations. Now consider the non-homogeneous equation (3.71).

Theorem 3.25. *If $W(x)$ is the fundamental matrix, while $\bar{y}(x)$ is a particular solution of the non-homogeneous equation (3.71), then any solution $y(x)$ of* *equation (3.71) may be represented in the form*

$$y(x) = W(x)C + \bar{y}(x), \qquad (3.85)$$

where C is a certain constant column.

The proof is exactly the same as in the case of n-th order equations; we shall omit it.

Let us construct a particular solution $\bar{y}(x)$ satisfying trivial initial conditions $\bar{y}(x_0) = 0$. We shall look for the solution in the form

$$\bar{y}(x) = W(x)C(x),$$

where $C(x)$ is an unknown column. This is in fact simply a change of variables. Substituting this in (3.71), we get

$$W'C + WC' = AWC + f.$$

Since W satisfies the homogeneous equation, it follows that $W' - AW = 0$ and therefore $WC' = f$. Hence $C' = W^{-1}f$. But since we had $\bar{y}(x_0) = W(x_0)C(x_0) = 0$, it follows that $C(x_0) = 0$ and therefore

$$C(x) = \int_{x_0}^{x} W^{-1}(\xi)f(\xi)d\xi.$$

Thus

$$\bar{y}(x) = \int_{x_0}^{x} W(x)W^{-1}(\xi)f(\xi)d\xi = \int_{x_0}^{x} \mathscr{K}(x,\xi)f(\xi)d\xi$$

and we have proved the following:

Theorem 3.26. *A particular solution $\bar{y}(x)$ of equation (3.71) satisfying the condition $\bar{y}(x_0) = 0$ is of the form*

$$\bar{y}(x) = \int_{x_0}^{x} \mathscr{K}(x,\xi)f(\xi)d\xi, \qquad (3.86)$$

where $\mathscr{K}(x,\xi)$ (the impulse matrix or matrizant) is the solution of the matrix equation (3.76) satisfying the condition $\mathscr{K}(\xi,\xi) = E$.

Remarks. 1. The method for constructing a particular solution developed above for systems of linear equations is in fact a version of the method of variation of constants which was used for one equation in Chapter 2.

2. In view of the superposition principle, the solution $y(x)$ of equation (3.71), (3.72) is of the form

$$y(x) = \mathscr{K}(x, x_0)y^0 + \int_{x_0}^{x} \mathscr{K}(x, \xi)f(\xi)d\xi. \tag{3.87}$$

§7. Systems of Linear Differential Equations with Constant Coefficients

Suppose that A in (3.74) is a constant matrix

$$y' = Ay, \quad A = \text{const.} \tag{3.88}$$

In this case the construction of the fundamental system of solutions or fundamental matrix reduces to algebraic operations.

We shall look for the particular solution of the system (3.88) in the form $\alpha e^{\lambda x}$, where λ is an unknown parameter and α an unknown constant column. Substituting this expression in (3.88), we obtain $\lambda\alpha e^{\lambda x} = A\alpha e^{\lambda x}$. Thus we conclude that α must be the solution of the algebraic system of equations

$$(A - \lambda E)\alpha = 0. \tag{3.89}$$

In order to obtain a non-trivial solution α, we must require

$$\text{Det } (A - \lambda E) = 0. \tag{3.90}$$

This equation is an algebraic equation of degree n and is called the characteristic equation of equation (3.88).

Suppose $\lambda_1, \ldots, \lambda_n$ are simple roots of the characteristic equation (3.90). To each λ_i corresponds an $\alpha_{(i)} \neq 0$ (the eigenvector of the matrix A), which can be found from (3.89), where we have put $\lambda = \lambda_i$. For the elements of $\alpha_{(i)}$ we can take, for example, the cofactors of the elements of one of the rows of the determinant Det $(A - \lambda E)$.

Theorem 3.27. *Suppose* $\lambda_1, \ldots, \lambda_n$ *are simple roots of the characteristic equation (3.90), and let* $\alpha_{(i)}$ *be a (non-trivial) solution of equation* $(A - \lambda_i E)\alpha = 0$. *Then the columns* $\alpha_{(i)}e^{\lambda_i x}$ ($i = 1, \ldots, n$) *constitute a fundamental system of solutions of equation (3.88).*

The proof will be carried out according to the outline used in §5. Assume that the solutions $\alpha_{(i)}e^{\lambda_i x}$ are linearly dependent

$$\sum_{i=1}^{n} C_i \alpha_{(i)} e^{\lambda_i x} = 0, \quad C_1 \neq 0. \tag{3.91}$$

This implies

$$C_1 \alpha_{(1)} e^{(\lambda_1 - \lambda_n)x} + \ldots + C_{n-1} \alpha_{(n-1)} e^{(\lambda_{n-1} - \lambda_n)x} + C_n \alpha_{(n)} = 0.$$

Taking derivatives in this relation, we obtain a relation of the same type as (3.91) but containing only $n - 1$ summands. Repeating this operation, we finally obtain the equality $C_1 \alpha_{(1)} = 0$. Since at least one of the elements of $\alpha_{(1)}$ is non-zero, we must conclude that $C_1 = 0$ which contradicts (3.91).

Let us consider the general case. Suppose the characteristic equation (3.90) possesses the roots $\lambda_1, \ldots, \lambda_l$ of multiplicities $m_1, \ldots, m_l (m_1 + \ldots + m_l = n)$. It follows from the previous arguments that $\alpha_{(i)} e^{\lambda_i x}$, where $\alpha_{(i)}$ is the eigenvector corresponding to λ_i, will be a solution of equation (3.88). In the case considered, more than one eigenvector may correspond to each λ_i, but, in general, their number is $p_i \leq m_i$. Thus there must be less than n solutions of the form $\alpha_{(i)} e^{\lambda_i x}$ and therefore they can no longer constitute a fundamental system of solutions.

In order to learn where to find the "missing" solutions, some constructions will be necessary. Namely, suppose y is a solution of equation (3.88). Then the components y_i of this solution satisfy the system of equations

$$a_{i1} y_1 + \ldots + a_{in} y_n - D y_i = 0 \quad (i = 1, \ldots, n), \tag{3.92}$$

where D is the differential operator (see remark 2 concerning Lemma 3.1). The determinant $\mathrm{Det}\,(A - ED) \equiv M(D)$ is a certain polynomial operator of the n-th degree. If instead of D we substitute λ, we obtain the left-hand side of the characteristic equation (3.90) or the characteristic polynomial of the system (3.88). Since multiplication of polynomial operators may be carried out according to the rules for multiplying ordinary polynomials, it follows that if we multiply (3.92) by the cofactor $A_{ij}(D)$ of the determinant $\mathrm{Det}\,(A - ED)$ (here multiplication is understood as the multiplication of operators) and take sums over i, we get

$$M(D) y_j = 0 \quad (j = 1, \ldots, n);$$

this is a differential equation of order n (with respect to y_j) whose characteristic polynomial coincides with the characteristic polynomial of the system (3.88). Thus we have proved the following:

Theorem 3.28. *Every component y_j of the solution of the system (3.88) satisfies an equation of the n-th order, whose characteristic polynomial equals the characteristic polynomial of the system (3.88).*

Consider the root λ_k of multiplicity m_k. The index k will be omitted in the following arguments, since we shall only deal with a single root. To this root λ corresponds the solution y of the system (3.88), whose j-th component y_j, according to Theorem 3.28, is of the form (see Theorem 3.14)

$$y_j = (C_{1j} + C_{2j}x + \ldots + C_{mj}x^{m-1})e^{\lambda x},$$

where the C_{kj} are constants, so that

$$y = \begin{pmatrix} C_{11} + C_{21}x + \ldots + C_{m1}x^{m-1} \\ \cdots\cdots\cdots\cdots\cdots\cdots\cdots\cdots \\ C_{1n} + C_{2n}x + \ldots + C_{mn}x^{m-1} \end{pmatrix} e^{\lambda x}. \qquad (3.93)$$

In this expression, however the constants C_{kj} are not independent, since the components y_j are not independent, being related by the system (3.92).

It turns out that the number of independent constants C_{kj} in expression (3.93) is equal to the multiplicity m of the root λ. We shall establish this fact later, and now we shall see what this yields for the construction of the fundamental system of solutions of Eq. (3.88).

Denote the free constants by $C_1, \ldots, C_m$. Substituting (3.93) into (3.88), dividing by $e^{\lambda x}$ and equalling the terms in identical powers of x to each other, we obtain a linear algebraic system of m homogeneous equations with $m \times n$ unknowns C_{kj} which may be expressed linearly in terms of the free constants $C_1, \ldots, C_m$. Then (3.93) may be rewritten in the form

$$y = [C_1 p_1(x) + \ldots + C_m p_m(x)]e^{\lambda x}, \qquad (3.94)$$

where the $p_i(x)$ are columns whose elements are well-defined polynomials with respect to x of degrees no higher than $m-1$.

If follows from (3.94) that m solutions of the form $p_i(x)e^{\lambda x}$ $(i=1, \ldots, m)$ correspond to the root λ of the characteristic equation of degree. This construction may be carried out for every λ_k of multiplicity m_k. The result will be $m_1 + \ldots + m_l = n$ solutions.

Further we shall prove that the n solutions obtained as above constitute a fundamental system of solutions of equation (3.88).

In practical terms, to find the fundamental system of solutions, it is recommended to write the expression (3.93) for every λ, then substitute it into (3.88) and express all the constants in terms of the free constants from the algebraic system thus obtained. The fact that the number of free constants is known in advance and equal to the multiplicity m of the root λ helps to solve this algebraic system, since this means that its rank is known in advance.

Example.

$$\begin{aligned} y_1' &= 4y_1 - y_2, \\ y_2' &= 3y_1 + y_2 - y_3, \\ y_3' &= y_1 + y_3. \end{aligned} \qquad (3.95)$$

The characteristic equation corresponding to this system has the root $\lambda_l = 2$ of multiplicity $m_1 = n = 3$. According to the rule described above, we write the expression

$$\begin{pmatrix} y_1 \\ y_2 \\ y_3 \end{pmatrix} = \begin{pmatrix} a_0 + a_1 x + a_2 x^2 \\ b_0 + b_1 x + b_2 x^2 \\ c_0 + c_1 x + c_2 x^2 \end{pmatrix} e^{2x}. \tag{3.96}$$

Substituting it into (3.95), dividing by e^{2x} and setting terms in identical powers of x equal to each other, we obtain the following 9 equations for determining 9 coefficients:

$$\begin{aligned}
a_1 &= 2a_0 - b_0, & 2a_2 &= 2a_1 - b_1, & 0 &= 2a_2 - b_2, \\
b_1 &= 3a_0 - b_0 - c_0, & 2b_2 &= 3a_1 - b_1 - c_1, & 0 &= 3a_2 - b_2 - c_2, \\
c_1 &= a_0 - c_0, & 2c_2 &= a_1 - c_1, & 0 &= a_2 - c_2.
\end{aligned} \tag{3.97}$$

It is known in advance that the rank of this system equals 6; so the number of free unknowns is 3.

If we write out the determinant of this system, arranging the unknowns in the order $a_0, b_0, c_0, a_1, b_1, c_1, a_2, b_2, c_2$, it is easy to see that the right upper determinant of the 6th order differs from zero and is therefore equal to the product of its diagonal elements, i.e. equal to 8, since to the right of the main diagonal there are only zeros. Therefore we may take a_0, b_0, c_0 for the free unknowns.

The first group of equations (3.97) already gives us expressions for a_1, b_1, c_1 in terms of a_0, b_0, c_0, and if we substitute this into the second group of equations (3.97), we obtain

$$a_2 = \frac{1}{2}(a_0 - b_0 + c_0), \; b_2 = a_0 - b_0 + c_0, \; c_2 = \frac{1}{2}(a_0 - b_0 + c_0).$$

The third group of equations (3.97) turns into an identity automatically.

Substituting the expressions obtained above into (3.96) and rewriting this in the form (3.94), we get

$$\begin{pmatrix} y_1 \\ y_2 \\ y_3 \end{pmatrix} = \left[a_0 \begin{pmatrix} 1 + 2x + \frac{1}{2} x^2 \\ 3x + x^2 \\ x + \frac{1}{2} x^2 \end{pmatrix} + b_0 \begin{pmatrix} -x - \frac{1}{2} x^2 \\ 1 - x - x^2 \\ -\frac{1}{2} x^2 \end{pmatrix} + c_0 \begin{pmatrix} \frac{1}{2} x^2 \\ -x - x^2 \\ 1 - x + \frac{1}{2} x^2 \end{pmatrix} \right] e^{2x}. \tag{3.98}$$

Here as a_0, b_0, c_0 are arbitrary constants, they may be denoted by C_1, C_2, C_3 as in (3.94), and the vectors $p_1(x), p_2(x), p_3(x)$ can be recognized in the right-hand side of (3.98). Thus we have obtained the solution of the system (3.88) in the form of a linear combination of three linearly independent solutions $p_i(x) e^{2x}$ $(i = 1, 2, 3)$.

In order to validate the methods used here, we must in fact establish two statements: first of all, show that the number of independent constants C_{kj} in (3.93) is equal to the multiplicity m of the root λ, and secondly, show that solutions of the form

$$p_{ki}(x)e^{\lambda_k x} \quad (i=1,\ldots,m_k; \; k=1,\ldots,l)$$

do indeed constitute a fundamental system of solutions of equation (3.88). To do this, we shall need a more precise idea of the structure of solutions corresponding to each root of the characteristic equation than the one given by formula (3.93). Let us now make this specification. Renumber the roots of the characteristic equations, or, equivalently, the characteristic roots (eigenvalues) of matrix A; namely, let us enumerate the eigenvectors. Then each value of λ will be enumerated a number of times equal to the number of linearly independent eigenvectors which correspond to it. For example, if to λ_1 correspond the eigenvectors $\alpha_{(11)}, \alpha_{(12)}, \ldots, \alpha_{(1\,p_1)}$ and to λ_2 the eigenvectors $\alpha_{(21)}, \alpha_{(22)}, \ldots, \alpha_{(2\,p_2)}$ etc., we shall say that we have the characteristic roots $\lambda_1, \lambda_2, \ldots, \lambda_{p_1}$, $\lambda_{p_1+1}, \ldots, \lambda_{p_2}$ etc. (here $\lambda_1 = \ldots = \lambda_{p_1}$, $\lambda_{p_1+1} = \ldots \lambda_{p_2}$ etc.). Thus we have $\lambda_1, \ldots, \lambda_s$ and to each of them corresponds an eigenvector.

The subsequent constructions are based on the following theorem, proved in linear algebra courses.

Theorem 3.29. *There exist n linearly independent constant vectors (columns) $e_{(kj_k)}(k=1,\ldots,s; \; j_k=1,\ldots,q_k)$ satisfying the relations*

$$
\begin{aligned}
Ae_{(k1)} &= \lambda_k e_{(k1)} \\
Ae_{(k2)} &= \lambda_k e_{(k2)} + e_{(k1)} \qquad\qquad (k=1,\ldots,s; \\
&\cdots\cdots\cdots\cdots\cdots\cdots\cdots\qquad q_1 + \ldots + q_s = n) \qquad (3.99) \\
Ae_{(kq_k)} &= \lambda_k e_{(kq_k)} + e_{(k\,q_k-1)}.
\end{aligned}
$$

so that the sum of q_k corresponding to the same λ_k equals m, where m is the multiplicity of the root λ_k of the characteristic equation (3.90).

In (3.99), $e_{(k1)}$ denotes the eigenvector corresponding to λ_k. The vectors $e_{(k2)}, \ldots, e_{(kq_k)}$ are called *adjoint vectors* generated by the eigenvectors $e_{(k1)}$. Thus to every λ_k correspond q_k linearly independent vectors, among which one is an eigenvector and the others are adjoint vectors, while to all the $\lambda_1, \ldots, \lambda_s$ correspond n linearly independent vectors. Recall that for different k the λ_k may be identical.

Consider λ_k. The solution $y_{(k1)} = e_{(k1)}e^{\lambda_k x}$ corresponds undoubtedly to it. It turns out that $q_k - 1$ more solutions correspond to it (so that there are q_k solutions in all), and this is shown by the following theorem.

Theorem 3.30. *To each λ_k correspond q_k solutions of the form*

$$y_{(k1)} = e_{(k1)} \exp \lambda_k x$$

$$y_{(k2)} = (e_{(k2)} + x e_{(k1)}) \exp (\lambda_k x)$$

$$\cdots\cdots\cdots\cdots\cdots\cdots\cdots\cdots\cdots\cdots\cdots \qquad (3.100)$$

$$y_{(kj)} = (e_{(kj)} + x e_{(kj-1)} + \ldots + \frac{x^{j-1}}{(j-1)!} e_{(k1)}) \exp (\lambda_k x)$$

$$y_{(kq_k)} = (e_{(kq_k)} + x e_{(kq_k-1)} + \ldots + \frac{x^{q_k-1}}{(q_k-1)!} e_{(k1)}) \exp (\lambda_k x).$$

This can easily be proved by a direct substitution, using (3.99). Indeed,

$$(A - ED) \left\{ e_{(kj)} + x e_{(kj-1)} + \ldots + \frac{x^{j-1}}{(j-1)!} e_{(k1)} \right\} \exp (\lambda_k x)$$

$$= (A - \lambda_k E) \left\{ \ldots \right\} - \left\{ e_{(kj-1)} + \ldots + \frac{x^{j-2}}{(j-2)!} e_{(k1)} \right\} \exp (\lambda_k x)$$

$$= \exp (\lambda_k x) \left\{ (A - \lambda_k E) e_{(kj)} + \ldots + \frac{x^{j-1}}{(j-1)!} (A - \lambda_k E) e_{(k1)} \right\}$$

$$- \left\{ e_{(kj-1)} + \ldots + \frac{x^{j-2}}{(j-2)!} e_{(k1)} \right\} \exp (\lambda_k x)$$

$$= \exp (\lambda_k x) \left\{ e_{(kj-1)} + \ldots + \frac{x^{j-2}}{(j-2)!} e_{(k1)} \right\}$$

$$- \left\{ e_{(kj-1)} + \ldots + \frac{x^{j-2}}{(j-2)!} e_{(k1)} \right\} \exp (\lambda_k x) = 0.$$

Thus,, to every $\lambda_k (k = 1, \ldots, s)$ correspond q_k solutions of the form (3.100), and therefore there are $q_1 + \ldots + q_s = n$ solutions in all:

$$y_{(11)}, \ldots, y_{(1q_1)}$$

$$\cdots\cdots\cdots\cdots \qquad (3.101)$$

$$y_{(s1)}, \ldots, y_{(sq_s)}.$$

Theorem 3.31. *The solutions (3.101) constitute a fundamental system of solutions.*

Indeed
$$y_{(k1)}(0) = e_{(k1)}, \ldots, y_{(kq_k)}(0) = e_{(kq_k)}$$

and, according to Theorem 3.29, the columns $e_{(k1)}, \ldots, e_{(kq_k)}$, $(k = 1, \ldots, s)$ whose number is $q_1 + \ldots + q_s = n$ are linearly independent and therefore

Det $W(0) \neq 0$. By Theorem 3.19, this means the solutions (3.101) are linearly independent, i.e. constitute a fundamental system of solutions.

Now let us return to the old numbering of the roots of the characteristic equation, when only different roots are assigned different numbers. To each λ there may now correspond several groups of solutions of the form (3.101) according to the number of eigenvectors corresponding to λ, but the total number of solutions in these groups is equal to the multiplicity m of the root λ. Thus the actual linear combination of solutions corresponding to the given λ is of the form (3.93), where there will be m independent constants, so that the number of solutions of the type (3.101) corresponding to this λ is m. Note that it can be seen from (3.100), (3.101) that the highest degree of polynomials in (3.93) is in general less than $m-1$.

In the practical calculation of a fundamental system of solutions we may use (3.100), first having calculated all the eigenvectors and the adjoint vectors, but it is easy to work as indicated above, substituting (3.93) into the given equation (3.88) and determining the m free unknowns C_{kj}.

§8. The Solutions in Power Series Form of Linear Equations

Linear equations with constant coefficients are a class of equations for which the fundamental system of solutions may be effectively written out.

How can we construct the fundamental system of solutions in the general case of an equation with variable coefficients? The aim of the present section is to give an idea of methods used to construct fundamental systems of solutions, based on the theory of power series and applicable in the case when the coefficients of the equations are analytic functions, i.e. may be represented as power series. The idea of the method is the following. In the simplest case, we write the solution in the form of a series $\sum\limits_{k=0}^{\infty} a_k x^k$, substitute it into the equation, in which the coefficients are also written in power series form; then the entire left-hand side will be written in power series form. Setting the coefficients of each power of x equal to zero, we obtain an equation to determine a_k. Note that when we construct the solution in power series form, in many cases we obtain the so-called special functions, which are widely used both in theoretical and in applied questions (e.g. Bessel functions, hypergeometric functions, etc.).

It is not our goal to develop the method in the general case – this is the object of a special branch of the theory of differential equations, known as the analytic theory of differential equations (see, for example [20], Chapter 5), and we shall demonstrate this method in the case of a simple equation often appearing in applications.

Consider the equation

$$y'' + xy = 0. \tag{3.102}$$

known as the Airy equation. It appears in various applications, for example, in quantum mechanics. This is the simplest equation of the second order with variable coefficient; however, it cannot be solved by elementary methods.

We shall look for the solution of equation (3.102) in the form of the series

$$y = \sum_{k=0}^{\infty} a_k x^k. \tag{3.103}$$

We have no *a priori* knowledge concerning the convergence of this series, and therefore all the operations which we shall now carry out with it will be formal in character; these operations must be viewed as an algorithm for determining the coefficients of the series for $y(x)$. Once the coefficients have been determined, we shall establish the convergence of the series (3.103) and the fact that the function $y(x)$ which it determines is indeed the solution of equation (3.102).

Thus, taking formal derivatives of the series (3.103) and substituting them into (3.102), we obtain

$$\sum_{k=2}^{\infty} a_k k(k-1)x^{k-2} + \sum_{k=0}^{\infty} a_k x^{k+1} = 0. \tag{3.104}$$

Now set the coefficients of identical powers of x equal to each other. We get

$$
\begin{aligned}
x^0) \quad & a_2 \cdot 2 \cdot 1 = 0; & \text{hence} \quad & a_2 = 0, \\
x^1) \quad & a_3 \cdot 3 \cdot 2 + a_0 = 0; & \text{hence} \quad & a_3 = -\frac{a_0}{3 \cdot 2}, \\
x^{k-2}) \quad & a_k \cdot k \cdot (k-1) + a_{k-3} = 0; & \text{hence} \quad & a_k = -\frac{a_{k-3}}{k(k-1)}.
\end{aligned}
\tag{3.105}
$$

It follows from (3.105) that
1) coefficients of the form a_{3q} can be expressed in terms of a_0:

$$a_{3q} = -\frac{1}{3q(3q-1)} a_{3q-3} = \ldots = \frac{(-1)^q}{3q \cdot (3q-1)\ldots 3 \cdot 2} a_0,$$

and the number a_0 remains undetermined,
2) the coefficients of the form a_{3q+1} can be expressed in terms of a_1:

$$a_{3q+1} = \frac{(-1)^q}{(3q+1) \cdot 3q \ldots 4 \cdot 3} a_1,$$

and the number a_1 itself remains undetermined, and finally,

3) the coefficients of the form a_{3q+2} can be expressed in terms of a_2:

$$a_{3q+2} = \frac{(-1)^q}{(3q+2)\cdot(3q+1)\ldots 5\cdot 4}\, a_2 = 0,$$

since $a_2 = 0$.

Put $a_0 = 1$, $a_1 = 0$. We obtain the series

$$y_1(x) = \sum_{q=0}^{\infty} a_{3q} x^{3q} = 1 + \sum_{q=1}^{\infty} \frac{(-1)^q}{3q\cdot(3q-1)\ldots 3\cdot 2}\, x^{3q}. \qquad (3.106)$$

Now conversely, putting $a_0 = 0$, $a_1 = 1$, we obtain the series

$$y_2(x) = \sum_{q=0}^{\infty} a_{3q+1} x^{3q+1} = x + \sum_{q=1}^{\infty} \frac{(-1)^q}{(3q+1)\cdot 3q\ldots 4\cdot 3}\, x^{3q+1}. \qquad (3.107)$$

Theorem 3.32. *The series (3.106) and (3.107) converge and the functions $y_1(x)$ and $y_2(x)$ which they determine constitute a fundamental system of solutions of equation (3.102).*

Indeed, the convergence of series (3.106) and (3.107) can be shown easily for every x, e.g. by using the d'Alembert rule, so that $y_1(x)$ and $y_2(x)$ are defined everywhere on $(-\infty, \infty)$. Since a power series may be differentiated any number of times on its interval of convergence, it follows that, after we take derivatives, substitute them into the left-hand side of (3.102) (see 3.104) and add both series, we obtain, according to the definition of coefficients a_k, the series consisting of zero terms. Therefore the equation becomes an identity.

The linear independence of $y_1(x)$ and $y_2(x)$ may be proved by *reductio ad absurdum*. Suppose $C_1 y_1(x) + C_2 y_2(x) \equiv 0$ and $C_1^2 + C_2^2 \neq 0$. Let $x = 0$. Then $y_2(0) = 0$, $y_1(0) = 1$ and therefore $C_1 = 0$. We then have $C_2 y_2(x) \equiv 0$, but then $y_2(x) \equiv 0$, which contradicts, for example, the fact that $y_2'(0) = 1$. Thus our theorem is proved.

Chapter IV
Boundary Value Problems

§1. Formulation of Boundary Value Problems and their Physical Meaning

In the previous chapters of our study of differential equations, we were mainly concerned with the solution of the Cauchy initial value problem, in which the supplementary conditions are initial ones: they determine the value of the unknown function and its derivatives for a fixed value of the independent variable. However, as we indicated in Chapter 1, initial conditions are not the only possible form of additional requirement specifying a definite particular solution. In many cases the additional requirements are given in the form of boundary conditions, determining the values of the unknown function and its derivatives (or certain expressions in the latter) for certain fixed values of the independent variable. The problem of finding a particular solution of the differential equation satisfying the given boundary conditions will be referred to as a boundary value problem. The study of the general properties and solution methods for boundary value problems constitutes the contents of the present chapter, the accent being placed on the study of boundary value problems for linear differential equations of the second order.

Many mathematical and physical problems reduce to boundary value problems for differential equations. Thus the problem of determining the static equilibrium position of an elastic rod with elasticity coefficient $k(x)$, fixed in its boundary sections and subjected to the action of an external force $f(x)$, considered in Chapter I, reduces to the boundary value problem

$$\frac{d}{dx}\left[k(x)\frac{du}{dx}\right] = -f(x), \quad u(0)=0, \quad u(l)=0. \tag{4.1}$$

Similarly, for the amplitude $u(x)$ of stable harmonic oscillations of frequency ω, we obtain the boundary value problem

$$\frac{d}{dx}\left[k(x)\frac{du}{dx}\right] + \omega^2 \varrho(x)u(x) = -f(x), \quad u(0)=0, \quad u(l)=0. \tag{4.2}$$

Here $\varrho(x)$ is the density of the rod. In the case when the values of the boundary section's displacement or the external forces acting on these sections are given, the boundary conditions must be replaced by non-homogeneous conditions of the form

$$u(0)=u_0, \text{ or } k(0)\frac{du}{dx}(0)=\vec{f}(0). \tag{4.3}$$

The problem of determining the trajectory of a particle of mass m which leaves the given point r_0 $(r(t_0)=r_0)$ at the initial moment of time t_0 and arrives at the point r_1 $(r(t_1)=r_1)$ at the moment t_1 reduces to the boundary value problem for the system of second order differential equations:

$$m\frac{d^2r}{dt^2}=F\left(t,r,\frac{dr}{dt}\right) \tag{4.4}$$

From here on, we will consider boundary value problems on the segment $[0,l]$ of the axis $0x$ for linear differential equations of the second order

$$y''+g(x)y'+h(x)y=f_1(x), \tag{4.5}$$

where $g(x), h(x), f_1(x)$ are continuous functions on $[0,l]$. Introduce the function

$$p(x)=e^{\int_0^x g(\xi)d\xi} \tag{4.6}$$

and note that

$$p(x)\frac{d^2y}{dx^2}+p(x)g(x)\frac{dy}{dx}=\frac{d}{dx}\left[p(x)\frac{dy}{dx}\right]. \tag{4.7}$$

Multiplying (4.5) by $p(x)$ (obviously $p(x)\neq 0$ on $[0,l]$), we obtain

$$L[y]\equiv\frac{d}{dx}\left[p(x)\frac{dy}{dx}\right]-q(x)y(x)=f(x), \tag{4.8}$$

where $q(x)=-p(x)h(x)$, $f(x)=p(x)f_1(x)$, while $L[y]$ denotes the differential operator in the left-hand side of (4.8).

Thus, without loss of generality, the study of boundary value problems for the general linear second order differential equation (4.5) may be reduced to the study of the same problem for Eq. (4.8).

Boundary value problems for Eq. (4.8), as a rule, are considered with linear boundary conditions of the form

$$\alpha_1 y'(0)+\beta_1 y(0)=u_0,$$
$$\alpha_2 y'(l)+\beta_2 y(l)=u_1, \tag{4.9}$$

(In (4.9) unilateral derivatives are understood.) Where α_i, β_i $(i=1, 2), u_0, u_l$ are given numbers, some of which may equal zero but $\alpha_i^2 + \beta_i^2 \neq 0$ (for both $i=1, 2$). If $\alpha_i = 0$ (for both $i=1, 2$), the corresponding boundary condition is usually called a condition of the *first kind*; if $\beta_i = 0$ (for both $i=1, 2$), it is called a condition of the *second kind*; if α_i and β_i are both different from zero it is a condition of the *third kind*.

Boundary value problems in which the right-hand side of the equation is non-zero will be called *non-homogeneous* boundary value problems. We consider these problems in §2 of the present chapter.

Boundary value problems for homogeneous equations with homogeneous boundary conditions will be called *homogeneous* boundary value problems.

It is obvious that the homogeneous boundary value problem, e.g. the problem

$$L[y]=0, \quad 0<x<l,$$
$$y(0)=0, \quad y(l)=0$$

always possesses an identically zero solution, called the trivial solution $y(x) \equiv 0$. It may turn out that the problem has no other solutions. This means that in solving the initial value problem for the equation $L[y]=0$ on the segment $0 \leq x \leq l$ with initial condition $y'(0)=y_1$ (in view of the existence theorem, the solution of the initial value problem (Theorem 3.1) exists and for $y_1 \neq 0$ is not identically zero) we shall obtain (for all values $y_1 \neq 0$) the function $y(x, y_1)$ possessing the property $y(l, y_1) \neq 0$. A similar state of the affairs arises when we set the corresponding initial values on the right extremity of the segment and construct the integral curves from right to left. For example, the problem

$$y''(x)=0, \ 0<x<1,$$
$$y(0)=y(1)=0$$

possesses only the trivial solution $y(x) \equiv 0$.

If for some value $\bar{y}_1 \neq 0$ the solution $y(x, \bar{y}_1)$ of the initial value problem constructed as above satisfies the second boundary condition $y(l, \bar{y}_1)=0$, for $x=l$ then this means that the given homogeneous boundary value problem also possesses, besides the trivial solution, a solution which is not identically zero. Such a solution of the homogeneous boundary value problem will be called non-trivial. Note that, since the problem is linear in this case, the function $Cy(x, \bar{y}_1)$ is also a non-trivial solution of the given homogeneous problem for any value of the constant $C \neq 0$.

For example, the problem

$$y''+y=0, \ 0<x<\pi,$$
$$y(0)=y(\pi)=0,$$

as can be easily checked, possesses, besides the trivial solution $y(x) \equiv 0$, the non-trivial solution $y(x) = C \sin x$.

The above arguments remain valid in the case of the more general boundary conditions (4.9).

An important particular case of homogeneous boundary value problems are the so-called eigenvalue problems; they consist in determining the parameters involved in the differential equation for which there exists a non-trivial solution of the homogeneous boundary value problem.

A typical eigenvalue problem for linear differential equations of the second order is the problem of finding values of the parameter λ for which there exists a non-trivial solution of the problem

$$L[y] + \lambda \varrho(x) y(x) = 0, \tag{4.10}$$

$$\begin{aligned} \alpha_1 y'(0) + \beta_1 y(0) &= 0, \\ \alpha_2 y'(l) + \beta_2 y(l) &= 0, \end{aligned} \tag{4.11}$$

on the segment $[0, l]$, where $L[y]$ is the same differential operator as in (4.8), $\varrho(x)$ is the given function, continuous on $[0, l]$, and $\alpha_i, \beta_i (i = 1, 2)$ are given constants, some of which may vanish.

Such problems as the determination of the natural oscillations of a material system with distributed characteristics (the transverse oscillations of a string, the transverse oscillations of an elastic rod, the acoustic oscillations in pipes, the electric oscillations in conductors, etc.) reduce to the eigenvalue problem (4.10)–(4.11). Suppose, for example, in the problem of harmonic oscillations of an elastic rod (see (4.2)), the right-hand side is equal to zero, i.e. there is no external force, while ω is a parameter. We are required to find out if harmonic oscillations of some frequency ω are possible in the absence of an external force (such oscillations are called free oscillations), i.e. if there exist values of ω for which the homogeneous problem (4.2) ($f(x) \equiv 0$) possesses a nontrivial solution.

Values of the parameter λ for which the problem (4.10)–(4.11) possesses a nontrivial solution are called *eigenvalues*, while the corresponding nontrivial solutions are the *eigenfunctions* of the boundary eigenvalue problem.

Remarks 1. Without loss of generality, the non-homogeneous boundary value problem may be considered with homogeneous boundary conditions. Indeed, in the case of non-homogeneous boundary conditions, we can look for the solution of the problem in the form

$$y(x) = Y(x) + z(x), \tag{4.12}$$

where the function $Y(x)$ satisfies only the given boundary conditions and is otherwise arbitrary. Obviously, such a function can always be constructed. For example, in the case of boundary conditions of the first kind $y(0)=u_0$, $y(l)=u_l$, for $Y(x)$ we can take the linear function

$$Y(x)=u_0\frac{l-x}{l}+u_l\frac{x}{l}.$$ (4.13)

Then we obtain a non-homogeneous boundary value problem for the function $z(x)$ with a somewhat different right-hand side, but with homogeneous boundary conditions.

2. In the general case of an n-th order equation, we must consider the boundary value problem with boundary conditions of more general form

$$P_i(y(0),\ldots,y^{(n-1)}(0),y(l),\ldots,y^{(n-1)}(l))=0,$$ (4.14)

where n is the order of the equation, while the $P_i(\cdot)$ are given functions of the solution and its derivatives at the boundary points. For example, conditions of type (4.14) are satisfied in the problem of finding periodic solutions when the supplementary conditions are of the form

$$y(0)=y(l),\ y'(0)=y'(l),\ldots,\ y^{(n-1)}(0)=y^{(n-1)}(l).$$ (4.15)

Let us note some properties of the solutions of Eq. (4.8) which will be important further on. Suppose $y(x)$ and $z(x)$ satisfy the equations

$$L[y]=\frac{d}{dx}\left[p(x)\frac{dy}{dx}\right]-q(x)y=f(x)$$ (4.16)

and

$$L[z]=\frac{d}{dx}\left[p(x)\frac{dz}{dy}\right]-q(x)z=g(x).$$ (4.17)

Multiplying (4.16) by $z(x)$, (4.17) by $y(x)$ and substracting term by term, we get

$$z(x)\frac{d}{dx}\left[p(x)\frac{dy}{dx}\right]-y(x)\frac{d}{dx}\left[p(x)\frac{dz}{dx}\right]=f(x)z(x)-g(x)y(x).$$ (4.18)

Since

$$\frac{d}{dx}\left[p(x)\left(z\frac{dy}{dx}-y\frac{dz}{dx}\right)\right]=z\frac{d}{dx}\left[p(x)\frac{dy}{dx}\right]-y\frac{d}{dx}\left[p(x)\frac{dz}{dx}\right]$$

$$+p(x)\left[\frac{dz}{dx}\frac{dy}{dx}-\frac{dy}{dx}\frac{dz}{dx}\right],$$

we see that (4.18) may be written in the form

$$\frac{d}{dx}\left[p(x)\left(z\frac{dy}{dx}-y\frac{dz}{dx}\right)\right]=f(x)z(x)-g(x)y(x). \tag{4.19}$$

This relation is known as the *Lagrange identity*. Its integral form is called *Green's formula*

$$\int_0^l (zL[y]-yL[z])dx=\left\{p(x)\left(z\frac{dy}{dx}-y\frac{dz}{dx}\right)\right\}\bigg|_0^l$$

$$=\int_0^l (f(x)z(x)-g(x)y(x))dx. \tag{4.20}$$

It follows from formula (4.19) that if $y_1(x)$ and $y_2(x)$ are two linearly independent solutions of the homogeneous equation $(f(x)\equiv g(x)\equiv 0)$

$$L[y]=\frac{d}{dx}\left[p(x)\frac{dy}{dx}\right]-q(x)y=0,$$

then they satisfy the relation

$$p(x)\left[y_1\frac{dy_2}{dx}-y_2\frac{dy_1}{dx}\right]=C, \tag{4.21}$$

which implies that the Wronskian of these solutions is of the form

$$\Delta(y_1,y_2)=\frac{C}{p(x)}. \tag{4.22}$$

Remarks. 1. Since the solutions of a homogeneous equation are determined up to an arbitrary constant factor, the constant C in (4.22) may be determined only by choosing the multiplicative factors of the solutions, i.e. by carrying out a so-called normalization of the solutions.

2. It follows from relation (4.22) that if we know some solution $y_1(x)$ of the homogeneous equation, then any other solution of this equation $y(x)$ which is linearly independent from $y_1(x)$ satisfies the first order linear differential equation

$$y_1(x)\frac{dy}{dx}-\frac{dy_1}{dx}y(x)=\frac{C}{p(x)}. \tag{4.23}$$

If $y_1(x)\neq 0$ on $[0,l]$ then, dividing (4.23) by $y_1^2(x)$, we obtain

$$\frac{d}{dx}\left(\frac{y(x)}{y_1(x)}\right)=\frac{C}{p(x)y_1^2(x)}, \tag{4.24}$$

which enables us to write the general solution of 4.23 in the form

$$y(x) = C_1 y_1(x) + C y_1(x) \int_0^x \frac{d\xi}{p(\xi) y_1^2(\xi)}.$$ (4.25)

§ 2. Non-homogeneous Boundary Value Problems

Consider the boundary value problem

$$L[y] = f(x),$$
$$\alpha_1 y'(0) + \beta_1 y(0) = 0,$$ (4.26)
$$\alpha_2 y'(l) + \beta_2 y(l) = 0.$$

Suppose the function $p(x)$ is positive and continuously differentiable on $[0, l]$, while the real-valued functions $q(x)$ and $f(x)$ are continuous on $[0, l]$. By a solution of the boundary value problem (4.26), we mean a function $y(x)$ continuously differentiable on $[0, l]$ with a continuous second derivative on $(0, l)$, satisfying the equation and the boundary conditions (4.26) on $(0, l)$. At first, we shall assume that the corresponding homogeneous boundary value problem possesses the trivial solution only. In other words, we assume that $\lambda = 0$ is not an eigenvalue of the corresponding problem (4.10)–(4.11).

Note at once that, since the problem (4.26) is linear, the previous assumption implies that *if the solution of the given problem exists, then it is unique.*

Our first objective is to prove the existence of the solution of problem (4.26) under the above assumptions concerning the coefficients and the right-hand side of the equation. The proof of the solution's existence will also contain an algorithm for its effective construction.

We begin with some helpful considerations. Assume that there exists a solution of the problem (4.26) when the right-hand side of the equation is given in a special form, namely when the function $f(x)$ differs from zero only in an ε-neighbourhood of some fixed point $x = \xi \in (0, l)$:

$$f(x) = \begin{cases} 0, & x \leq \xi - \varepsilon, \\ f_\varepsilon(x), & \xi - \varepsilon \leq x \leq \xi + \varepsilon, \\ 0, & x \geq \xi + \varepsilon, \end{cases}$$ (4.27)

the function $f_\varepsilon(x)$ being non-negative and

$$\int_{\xi-\varepsilon}^{\xi+\varepsilon} f_\varepsilon(x) dx = 1.$$ (4.28)

The solution of this problem will be denoted by $y_\varepsilon(x, \xi)$.

Integrating equation (4.26) for such a function $f(x)$ over the segment $[\xi - \varepsilon, \xi + \varepsilon]$, we obtain

$$p(\xi + \varepsilon) y'_\varepsilon(\xi + \varepsilon, \xi) - p(\xi - \varepsilon) y'_\varepsilon(\xi - \varepsilon, \xi)$$

$$- \int_{\xi - \varepsilon}^{\xi + \varepsilon} q(x) y_\varepsilon(x, \xi) dx = \int_{\xi - \varepsilon}^{\xi + \varepsilon} f_\varepsilon(x) dx = 1. \qquad (4.29)$$

Now consider the passage to the limit as $\varepsilon \to 0$, assuming that (4.29) is valid for any ε and therefore

$$\lim_{\varepsilon \to 0} \int_{\xi - \varepsilon}^{\xi + \varepsilon} f_\varepsilon(x) dx = 1. \qquad (4.30)$$

We shall also assume that the limit function

$$\lim_{\varepsilon \to 0} y_\varepsilon(x, \xi) = G(x, \xi) \qquad (4.31)$$

exists and is continuous on $[0, l]$. Then, passing to the limit for $\varepsilon \to 0$ in (4.29), we see that the derivative $\dfrac{d}{dx} G(x, \xi)$ at the point $x = \xi$ must have a discontinuity of the first kind, while the difference between the limiting values from the right and from the left of this derivative at the point $x = \xi$ is determined by the expression

$$\frac{d}{dx} G(x, \xi) \bigg|_{x = \xi + 0} - \frac{d}{dx} G(x, \xi) \bigg|_{x = \xi - 0} = \frac{1}{p(\xi)}. \qquad (4.32)$$

Summarizing the above, we can say that if the function $G(x, \xi)$ exists, then it must verify the following conditions:

1) as a function of the variable x, $G(x, \xi)$ satisfies the homogeneous equation (4.26) for $0 < x < \xi$ and $\xi < x < l$

2) $G(x, \xi)$ satisfies the boundary conditions (4.26);

3) $G(x, \xi)$ is continuous on $[0, l]$ and its first derivative possesses a discontinuity of the first kind at the point $x = \xi$ with a difference of limiting values equal to

$$\frac{d}{dx} G(x, \xi) \bigg|_{\xi - 0}^{\xi + 0} = \frac{1}{p(\xi)}. \qquad (4.32')$$

A function satisfying conditions 1), 2), 3) will be called *Green's function for the boundary value problem* (4.26). The importance of Green's function is due, in particular, to the fact that *the solution of the boundary value problem* (4.26) *with arbitrary right-hand side* $f(x)$ *can always be expressed in terms of Green's function.*

Indeed, suppose a solution of problem (4.26) exists and Green's function is $G(x, \xi)$. Applying Green's formula (4.20) to these functions on the segments $[0, \xi - \varepsilon]$ and $[\xi + \varepsilon, l]$ where the functions $y(x)$ and $G(x, \xi)$ are continuously differentiable and possess continuous second derivatives, we obtain

$$\left\{ p(x) \left(G(x, \xi) \frac{dy}{dx} - y(x) \frac{dG}{dx} \right) \right\} \Big|_0^{\xi - \varepsilon} + \left\{ p(x) \left(G(x, \xi) \frac{dy}{dx} - y(x) \frac{dG}{dx} \right) \right\} \Big|_{\xi + \varepsilon}^{l}$$

$$= \int_0^{\xi - \varepsilon} G(x, \xi) f(x) dx + \int_{\xi + \varepsilon}^{l} G(x, \xi) f(x) dx. \qquad (4.33)$$

Since $y(x)$, as well as $G(x, \xi)$, satisfy the homogeneous boundary conditions (4.26), the substitutions for $x = 0$ and $x = l$ vanish. Passing to the limit as $\varepsilon \to 0$ in (4.33), by definition of Green's function, we get

$$y(\xi) = \int_0^l G(x, \xi) f(x) dx, \qquad (4.34)$$

which proves the statement claimed above.

Now let us show that *Green's function for the boundary value problem (4.26) exists* and exhibit an algorithm for constructing it. Consider the function $y_1(x)$ which is the solution of the homogeneous equation (4.26) satisfying the left boundary condition

$$\alpha_1 y_1'(0) + \beta_1 y_1(0) = 0. \qquad (4.35)$$

It is obvious that under the given assumptions concerning the coefficients of the equation, the function $y_1(x)$ can always be constructed as the solution of the initial value problem for equation (4.26) with initial conditions

$$y_1(0) = -C\alpha_1, \qquad y_1'(0) = C\beta_1, \qquad (4.36)$$

where C is an arbitrary constant. Under our assumptions (on the existence of only trivial solutions of the homogeneous boundary value problem), the function $y_1(x)$ constructed above does not satisfy the right boundary condition

$$\alpha_2 y_1'(l) + \beta_2 y_1(l) \neq 0. \qquad (4.37)$$

Similarly, we can construct a function $y_2(x)$ which is the solution of the homogeneous equation satisfying the right boundary condition

$$\alpha_2 y_2'(l) + \beta_2 y_2(l) = 0. \qquad (4.38)$$

It is easy to see that the functions $y_1(x)$ and $y_2(x)$ constructed in this way are linearly independent. Indeed, if we assume that these functions are linearly dependent, e.g.

$$y_1(x) = C y_2(x), \tag{4.39}$$

we see that $y_1(x)$ satisfies the right boundary condition, which contradicts (4.37). Thus for the homogeneous equation (4.26) we have constructed two linearly independent solutions each of which satisfies only one of the two homogeneous boundary conditions.

Let us look for the function $G(x, \xi)$ in the form

$$G(x, \xi) = \begin{cases} C_1 y_1(x), & 0 \le x \le \xi, \\ C_2 y_2(x), & \xi \le x \le l. \end{cases} \tag{4.40}$$

It is then clear that the function (4.40) will satisfy the homogeneous equation (4.26) for $x \neq \xi$ and the homogeneous boundary conditions. In order to satisfy the continuity condition on $G(x, \xi)$ and the condition on the discontinuity of its derivative (4.32′), it remains only to find the constants C_1 and C_2 from the relations

$$C_2 y_2(\xi) - C_1 y_1(\xi) = 0, \quad C_2 y_2'(\xi) - C_1 y_1'(\xi) = \frac{1}{p(\xi)}. \tag{4.41}$$

The determinant of this linear algebraic system, which is in fact the Wronskian of the linearly independent solutions $y_1(x)$ and $y_2(x)$, is nonzero and by (4.22) we have

$$\Delta(y_1, y_2) = \frac{C}{p(x)}, \tag{4.22}$$

where the constant C is determined by normalizing the solutions $y_1(x)$ and $y_2(x)$. This implies that the system (4.41) can be solved uniquely. Substituting the obtained values of C_1 and C_2 into (4.40), we get the final expression for Green's function for the boundary value problem (4.26):

$$G(x, \xi) = \frac{1}{C} \begin{cases} y_2(\xi) y_1(x), & 0 \le x \le \xi, \\ y_1(\xi) y_2(x), & \xi \le x \le l, \end{cases} \tag{4.42}$$

where the constant equals

$$C = p(\xi) \Delta(y_1(\xi), y_2(\xi)). \tag{4.43}$$

The arguments carried out above prove the existence and uniqueness of Green's function for the boundary value problem (4.26) in the case when the corresponding homogeneous problem possesses only the trivial solution.

Remarks. 1. It follows from the explicit form (4.42) of Green's function that it is a *symmetric function* of its arguments

$$G(x, \xi) = G(\xi, x). \tag{4.44}$$

2. It follows from the introductory remarks preceding the construction of Green's function that it has a simple physical meaning, being the solution of the boundary value problem in the case of a concentrated load.

Example. As the simplest illustration, consider the construction of Green's function for the boundary value problem

$$y'' = f(x), \quad 0 < x < l$$
$$y(0) = 0, \quad y'(l) = 0.$$

It is easy to check directly that the corresponding homogeneous problem possesses only the trivial solution.

For the functions $y_1(x)$ and $y_2(x)$ satisfying, respectively, the left and the right boundary conditions, choose $y_1(x) = x$ and $y_2(x) = 1$. According to (4.40), we will look for Green's function for the problem under consideration in the form

$$G(x, \xi) = \begin{cases} C_1 x, & 0 \le x \le \xi, \\ C_2, & \xi \le x \le l. \end{cases}$$

In order to satisfy the continuity condition of the function $G(x, \xi)$ for $x = \xi$ and the requirement concerning the size of the discontinuity of the derivative at this point, we must put

$$C_1 \xi = C_2 \quad -C_1 = 1.$$

Hence

$$C_1 = -1, \quad C_2 = -\xi$$

and finally the expression of Green's function for the problem under consideration will be

$$G(x, \xi) = -\begin{cases} x, & 0 \le x \le \xi, \\ \xi, & \xi \le x \le l. \end{cases}$$

Note that according to (4.1) the Green's function constructed above determines the form of static equilibrium under the action of an exterior longitudinal concentrated force on a uniform elastic rod whose left extremity is fixed while the right extremity is free. Green's function $G(x, \xi)$ expresses the shift of the section x if a concentrated force contracting the rod is applied at the point ξ. As we can see from the expression for $G(x, \xi)$, under the given conditions, the shift of the rod's sections situated to the left of the point where the force is applied $(x \le \xi)$ will be non-uniform and will be greater when the given section is nearer to the point ξ where the force is applied; the rest of the rod, situated to the right of the point where the force is applied, will be shifted over as a solid body.

Now let us consider the existence of solutions of the given boundary value problem. We have the following.

Theorem 4.1. *If the homogeneous boundary value problem (4.26) possesses the trivial solution only then the solution of the non-homogeneous boundary value problem exists for any function $f(x)$ continuous on $[0, l]$ and is expressed by the formula*

$$y(x) = \int_0^l G(x, \xi) f(\xi) d\xi. \tag{4.45}$$

In order to prove the theorem, it is sufficient to verify directly that the function $y(x)$ given by formula (4.45) satisfies all the conditions in the definition of a solution of the boundary value problem (4.26).

Indeed, write formula (4.45) in the form

$$y(x) = \int_0^x G_1(x, \xi) f(\xi) d\xi + \int_x^l G_2(x, \xi) f(\xi) d\xi,$$

where $G_1(x, \xi) = \dfrac{1}{C} y_1(\xi) y_2(x)$, $G_2(x, \xi) = \dfrac{1}{C} y_2(\xi) y_1(x)$ are functions of x and ξ continuous together with their second order derivatives. The continuity condition of the function $G(x, \xi)$ at the point $x = \xi$ may be written in the form $G_1(\xi, \xi) = G_2(\xi, \xi)$ or $G_1(x, x) = G_2(x, x)$ while the assumption on the discontinuity of the derivative (4.32′) may be written in the form $G'_{1x}(x, x) - G'_{2x}(x, x)$ $= \dfrac{1}{p(x)}$.

Using the above relations between G_1 and G_2 and taking derivatives of the integrals in accordance to well known rules, we obtain

$$y'(x) = G_1(x, x) f(x) + \int_0^x G'_{1x}(x, \xi) f(\xi) d\xi - G_2(x, x) f(x)$$

$$+ \int_x^l G'_{2x}(x, \xi) f(\xi) f(\xi) d\xi = \int_0^x G'_{1x}(x, \xi) f(\xi) d\xi + \int_x^l G'_{2x}(x, \xi) f(\xi) d\xi,$$

$$y''(x) = G'_{1x}(x, x) f(x) + \int_0^x G''_{1xx}(x, \xi) f(\xi) d\xi$$

$$- G'_{2x}(x, x) f(x) + \int_x^l G''_{2xx}(x, \xi) f(\xi) d\xi = \frac{f(x)}{p(x)}$$

$$+ \int_0^x G''_{1xx}(x, \xi) f(\xi) d\xi + \int_x^l G''_{2xx}(x, \xi) f(\xi) d\xi.$$

Hence

$$L[y(x)] = py'' + p'y' - qy = \int_0^x L[G_1] f(\xi) d\xi + \int_x^l L[G_2] f(\xi) d\xi + f(x) = f(x),$$

which proves the statement of the theorem.

Now let us consider the case when the homogeneous boundary value problem (4.26) possesses a nontrivial solution. To simplify the computations which follow, we will consider the boundary value problem with boundary conditions of the first kind

$$L[y]=f(x), \quad y(0)=0, \quad y(l)=0. \tag{4.46}$$

By assumption, there exists a function $\varphi_0(x)$ which is the solution of the corresponding homogeneous boundary value problem

$$L[\varphi_0(x)]=0, \quad \varphi_0(0)=0, \quad \varphi_0(l)=0. \tag{4.47}$$

The function $\varphi_0(x)$ may be viewed as an eigenfunction of the problem (4.10)–(4.11) corresponding to the eigenvalue $\lambda=0$ when $\alpha_1 = \alpha_2 = 0$. Further we shall assume that the homogeneous boundary value problem (4.47) has no other solutions which are linearly independent from $\varphi_0(x)$. We normalize the function $\varphi_0(x)$ by setting

$$\int_0^l \varphi_0^2(x)dx=1. \tag{4.48}$$

Note at once that if the solution $y(x)$ of the boundary value problem (4.46) exists, then the right-hand side of the equation (the function $f(x)$) must be orthogonal on $[0, l]$ to the function $\varphi_0(x)$. Indeed, applying Green's formula (4.20) to the functions $y(x)$ and $\varphi_0(x)$, we get

$$\int_0^l (\varphi_0(x)L[y]-y(x)L[\varphi_0])dx=\int_0^l \varphi_0(x)f(x)dx$$

$$= [p(x)(\varphi_0(x)y'(x)-y(x)\varphi_0'(x))]|_0^l=0, \tag{4.49}$$

which implies our statement.

In other words, we have the following.

Lemma 4.1. *A necessary condition for the existence of the solution of the boundary value problem (4.46) is the orthogonality of the right-hand side of equation (4.46) to the nontrivial solution of the corresponding homogeneous boundary value problem (4.47).*

In the sequel we shall assume that the necessary condition for the existence of the solution of the boundary value problem 4.46 holds. Let us show that in this case the solution of the boundary value problem exists and may be constructed by means of an appropriately determined Green's function. Note that if the function $y(x)$ is a solution of the boundary value problem (4.46), then any function $\tilde{y}(x)$ of the form

$$\tilde{y}(x)=y(x)+C\varphi_0(x), \tag{4.50}$$

where C is an arbitrary constant, is a solution of the same problem, since the function $\varphi_0(x)$ is a solution of the corresponding homogeneous problem. Therefore, in order to specify a unique solution of the boundary value problem, it must be subjected to a supplementary condition. For this we choose the orthogonality of the unknown solution to the eigenfunction $\varphi_0(x)$:

$$\int_0^l y(x)\varphi_0(x)dx = 0. \tag{4.51}$$

Let us show that the problem (4.46), (4.51) may possess only one solution. From the linearity of the problem, it is obvious that we need only show that the corresponding homogeneous problem possesses the trivial solution only.

Lemma 4.2. *The homogeneous problem (4.47), (4.51) possesses the trivial solution only.*

By assumption, all the solutions of the homogeneous problem (4.47) can be represented in the form $y_0(x) = C\varphi_0(x)$. Condition (4.51) yields

$$\int_0^l y_0(x)\varphi_0(x)dx = C \int_0^l \varphi_0^2(x)dx,$$

so that $C=0$ and the lemma is proved.

We now turn to the construction of so-called *generalized Green's functions for the problem (4.46), (4.51)*. Since there exists a nontrivial solution of the homogeneous boundary value problem (4.47), in order to construct an appropriate Green's function, we cannot limit ourselves to the solutions of the homogeneous equation only. Therefore we shall define the generalized Green's function $G(x, \xi)$ as the solution of the following problem:

1. $G(x, \xi)$ satisfies the nonhomogeneous equation

$$L_x[G(x, \xi)] = -\varphi_0(\xi)\varphi_0(x) \tag{4.52}$$

for $0 < x < \xi$ and $\xi < x < l$.

2. $G(x, \xi)$ satisfies the same boundary conditions as the required solution

$$G(0, \xi) = G(l, \xi) = 0. \tag{4.53}$$

3. $G(x, \xi)$ is continuous on $[0, l]$.

4. The first derivative $\dfrac{d}{dx} G(x, \xi)$ possesses a discontinuity of the first kind for $x = \xi$ and the size of this discontinuity is

$$\frac{d}{dx} G(x, \xi) \big|_{\xi-0}^{\xi+0} = \frac{1}{p(\xi)}. \tag{4.54}$$

5. $G(x, \xi)$ is orthogonal on $[0, l]$ to the eigenfunction $\varphi_0(x)$:

$$\int_0^l G(x, \xi)\varphi_0(x)dx = 0. \qquad (4.55)$$

Let us show at once that if the solution of the given boundary value problem (4.46), (4.51) and the generalized Green's function exist, then the solution of the problem (4.46), (4.51) can be expressed in terms of the generalized Green's function according to a formula similar to (4.34). Indeed, applying Green's formula to the functions $y(x)$ and $G(x, \xi)$ on $[0, \xi - \varepsilon]$ and $[\xi + \varepsilon, l]$ and, passing to the limit as $\varepsilon \to 0$, we obtain as before

$$p(\xi)y(\xi)\left.\frac{dG}{dx}\right|_{x=\xi-0}^{x=\xi+0} = \int_0^l \{G(x, \xi)f(x) + y(x)\varphi_0(\xi)\varphi_0(x)\}dx, \qquad (4.56)$$

which, in view of conditions (4.54) and (4.51) finally yields

$$y(\xi) = \int_0^l G(x, \xi)f(x)dx. \qquad (4.57)$$

We shall indicate an algorithm for the explicit construction of the generalized Green's function determined by conditions 1–5. Thus we will also obtain a constructive proof of its existence.

Let us choose some particular solution $\omega(x)$ of the nonhomogeneous equation

$$L[\omega(x)] = -\varphi_0(\xi)\varphi_0(x) \qquad (4.58)$$

on $(0, l)$ (for example, the solution of the initial value problem for equation (4.58) with initial conditions $\omega(0) = \alpha$, $\omega'(0) = \beta$).

Together with the function $\varphi_0(x)$, consider the linearly independent solution $\varphi_1(x)$ of the homogeneous equation (4.46):

$$L[\varphi_1(x)] = 0, \qquad (4.59)$$

where we assume that $\varphi_1(x)$ is normalized so as to have

$$\Delta(\varphi_1(x), \varphi_0(x)) = \frac{1}{p(x)}. \qquad (4.60)$$

Note at once that the functions $\varphi_0(x)$ and $\varphi_1(x)$ are linearly independent, hence the function $\varphi_1(x)$ cannot satisfy the same homogeneous boundary conditions as $\varphi_0(x)$:

$$\varphi_1(0) \neq 0, \qquad \varphi_1(l) \neq 0. \qquad (4.61)$$

The particular solution $\omega(x)$ chosen above does not in general satisfy the homogeneous boundary conditions (4.53). However, it is possible to choose a linear combination of the functions $\omega(x)$ and $\varphi_1(x)$ on the segments $[0, \xi]$ and $[\xi, l]$ so as to satisfy conditions (4.53). But this does not leave sufficient freedom to satisfy the remaining conditions 3–5. To do this, we will add a solution of the homogeneous equation, linearly independent from φ_1 – namely, the function φ_0 which satisfies the homogeneous conditions (4.53). Hence the generalized Green's function may be constructed in the form

$$G(x, \xi) = \omega(x) + \begin{cases} C_1\varphi_1(x) + C_3\varphi_0(x), & 0 \le x \le \xi, \\ C_2\varphi_1(x) + C_4\varphi_0(x), & \xi \le x \le l, \end{cases} \tag{4.62}$$

by choosing the constants C_1, C_2, C_3, C_4 so as to satisfy all the conditions. Conditions (4.53) give us

$$\omega(0) + C_1\varphi_1(0) = 0, \quad \omega(l) + C_2\varphi_1(l) = 0. \tag{4.63}$$

By (4.61), this enables us to determine uniquely the constants C_1 and C_2. The continuity condition of the function $G(x, \xi)$ and the discontinuity condition of its derivative for $x = \xi$ yield the equations

$$(C_2 - C_1)\varphi_1(\xi) + (C_4 - C_3)\varphi_0(\xi) = 0,$$
$$(C_2 - C_1)\varphi_1'(\xi) + (C_4 - C_3)\varphi_0'(\xi) = \frac{1}{p(\xi)}. \tag{4.64}$$

By (4.60), this system possesses a unique solution with respect to $C_2 - C_1$ and $C_4 - C_3$:

$$C_2 - C_1 = -\varphi_0(\xi), \quad C_4 - C_3 = \varphi_1(\xi). \tag{4.65}$$

The values C_2 and C_1 were already found from (4.63). Let us show that they satisfy (4.65). Applying Green's formula to the functions $\omega(x)$ and $\varphi_0(x)$, we get

$$\int_0^l (\varphi_0(x)L[\omega] - \omega(x)L[\varphi_0])dx$$
$$= \{p(x)(\varphi_0(x)\omega'(x) - \omega(x)\varphi_0'(x))\}|_0^l = -\int_0^l \varphi_0(x)\varphi_0(\xi)\varphi_0(x)dx;$$

obtaining

$$-p(l)\omega(l)\varphi_0'(l) + p(0)\omega(0)\varphi_0'(0) = -\varphi_0(\xi). \tag{4.66}$$

But by (4.60),

$$p(l)\varphi_0'(l) = \frac{1}{\varphi_1(l)}, \quad p(0)\varphi_0'(0) = \frac{1}{\varphi_1(0)} \tag{4.67}$$

and, using (4.63), we can transform formula (4.66) into the expression $C_2 - C_1 = -\varphi_0(x)$ which coincides with (4.65). Thus C_1 and C_2 are uniquely determined. Let us express C_4 in terms of C_1, using (4.63)

$$C_4 = C_3 + \varphi_1(\xi) \tag{4.68}$$

and then rewrite (4.62) in the form

$$G(x,\xi) = \omega(x) + C_3\varphi_0(x) + \begin{cases} C_1\varphi_1(x), & 0 \le x \le \xi, \\ C_2\varphi_1(x) + \varphi_1(\xi)\varphi_0(x), & \xi \le x \le l. \end{cases} \tag{4.69}$$

Now only the coefficient C_3 is unknown. Substituting (4.69) into condition (4.55), we obtain the following expression for determining the coefficient C_3:

$$C_3 \int_0^l \varphi_0^2(x)dx = C_3 = -\left\{ \int_0^l \omega(x)\varphi_0(x)dx \right.$$
$$\left. + C_1 \int_0^\xi \varphi_1(x)\varphi_0(x)dx + C_2 \int_\xi^l \varphi_1(x)\varphi_0(x)dx + \varphi_1(\xi)\int_\xi^l \varphi_0^2(x)dx \right\}. \tag{4.70}$$

The coefficient C_4 is expressed in terms of C_3 according to formula (4.68). Thus the generalized Green's function is constructed.

The above arguments enable us to state the following theorem.

Theorem 4.2. *A necessary and sufficient condition for the existence of a unique solution of the boundary value problem (4.46), (4.51) is the orthogonality condition (4.49) of the right-hand side $f(x)$ of equation (4.46) to the eigenfunction $\varphi_0(x)$. When this condition holds, the solution may be expressed in the form*

$$y(x) = \int_0^l G(x,\xi)f(\xi)d\xi. \tag{4.71}$$

The necessity of the orthogonality condition (4.49) and the uniqueness of the solution of the boundary value problem were proved previously (see Lemmas 4.1 and 4.2); the proof of the fact that the function $y(x)$ determined by formula (4.71) under the assumption (4.49) is indeed a solution of the boundary value problem is a straightforward verification.

Remarks. 1. In order to simplify the calculations, we considered the boundary value problem with boundary conditions of the first kind. Similar considerations and constructions of the generalized Green's function are valid in the case of general boundary conditions.

2. We assumed that the homogeneous boundary value problem (4.47) possesses only one linearly independent solution $\varphi_0(x)$. A similar

argument may also be used in the case when this problem possesses two linearly independent solutions $\varphi_1(x)$, $\varphi_2(x)$ (the number of linearly independent solutions, obviously, is no greater than two, since the order of the equation equals two). In this case two linearly independent eigenfunctions $\varphi_1(x)$, $\varphi_2(x)$ correspond to the eigenvalue $\lambda=0$ of the boundary eigenvalue problem (4.10)–(4.11).

Here, the nonhomogeneous boundary value problem has a solution if we have the orthogonality condition of the right-hand side of the equation to both of the eigenfunctions:

$$\int_0^l f(x)\varphi_i(x)dx=0 \quad (i=1,2). \tag{4.72}$$

In order to have uniqueness, the solution of the boundary value problem must be subjected to similar orthogonality conditions to all the eigenfunctions, while the generalized Green's function is constructed as the solution of a non-homogeneous equation, from the right-hand side of which a linear combination of eigenfunctions is chosen.

Example. Consider the boundary value problem

$$y''+y=f(x), \quad 0<x<\pi$$
$$y(0)=0, \quad y(\pi)=0.$$

As we can easily see, the corresponding homogeneous boundary value problem possesses the nontrivial solution $\varphi_0(x)=\sqrt{\dfrac{2}{\pi}}\sin x$ (the coefficient $\sqrt{\dfrac{2}{\pi}}$ is chosen from the normalization condition (4.48). Hence the solution of the non-homogeneous boundary value problem exists only in the case when the right-hand side $f(x)$ satisfies the orthogonality condition (4.49)

$$\int_0^\pi f(x) \sin x dx=0$$

and, in order to represent the solution by formula (4.57), we need the generalized Green's function.

Let us construct it. To find a particular solution $\omega(x)$ of the equation

$$\omega''+\omega=-\frac{2}{\pi} \sin \xi \sin x$$

let us look for it in the form $\omega=Ax \cos x$. By substituting directly into the equation, we readily find the value of the constant $A=\dfrac{1}{\pi} \sin \xi$ and obtain

$$\omega(x)=\frac{1}{\pi} \sin \xi \, (x \cos x).$$

For our particular solution (of the homogeneous equation) which does not satisfy the given boundary conditions, we choose the function $\varphi_1(x)=\cos x$.

According to (4.62), the generalized Green's function of the given problem should be sought in the form

$$G(x,\xi)=\frac{1}{\pi}\sin\xi\,(x\cos x)+\begin{cases}C_1\cos x+C_3\sin x, & 0\le x\le\xi \\ C_2\cos x+C_4\sin x, & \xi\le x\le\pi.\end{cases}$$

The boundary conditions $G(0,\xi)=0$, $G(\pi,\xi)=0$ yield an equation for determining the constants C_1 and C_2

$$C_1=0,\quad -\sin\xi-C_2=0,$$

from which we get

$$C_1=0,\quad C_2=-\sin\xi.$$

The continuity condition of the function $G(x,\xi)$ for $x=\xi$ and the discontinuity of its derivative at this point, taking into consideration the values of C_1 and C_2 previously found, give us equations for finding the constants C_3 and C_4

$$C_3\sin\xi+\sin\xi\cos\xi-C_4\sin\xi=0,$$
$$\sin^2\xi+C_4\cos\xi-C_3\cos\xi=1,$$

hence

$$C_4=C_3+\cos\xi$$

and the expression for $G(x,\xi)$ acquires the form

$$G(x,\xi)=\frac{1}{\pi}\sin\xi\,(x\cos x)+C_3\sin x+\begin{cases}0 & 0\le x\le\xi \\ -\sin\xi\cos x+\cos\xi\sin x, & \xi\le x\le\pi\end{cases}$$

The constant C_3 may be found from the orthogonality condition of the function $G(x,\xi)$ to the function $\varphi_0(x)=\sqrt{\frac{2}{\pi}}\sin x$

$$\int_0^\pi G(x,\xi)\sin x\,dx=0.$$

Computing the appropriate integrals, we get

$$C_3=-\frac{\pi-\xi}{\pi}\cos\xi-\frac{1}{2\pi}\sin\xi,$$

which gives the final expression of the generalized Green's function for the given problem in the form

$$G(x,\xi) = \frac{1}{\pi}(x \sin \xi \cos x + \xi \cos \xi \sin x) - \frac{1}{2\pi} \sin \xi \sin x$$

$$- \begin{cases} \cos \xi \sin x, & 0 \le x \le \xi \\ \sin \xi \cos x, & \xi \le x \le \pi. \end{cases}$$

By using the above expression of the generalized Green's function and the representation (4.57), it is easy to obtain the solution of the non-homogeneous boundary value problem in the case when the right-hand side $f(x)$ of the equation satisfies the orthogonality condition

$$\int_0^\pi f(x)\varphi_0(x)dx = 0.$$

Let us choose the function $f(x)$ in a specific way, namely by putting $f(x) = \cos x$ and then try to find the solution of the boundary value problem

$$y'' + y = \cos x, \quad 0 < x < \pi$$
$$y(0) = 0, \quad y(\pi) = 0,$$

assuming that this solution is orthogonal to the eigenfunction $\sin x$ of the corresponding homogeneous boundary value problem

$$\int_0^\pi y(x) \sin x dx = 0.$$

Using the expression for the generalized Green's function just obtained above and the representation (4.57), after simple calculations of elementary integrals we get

$$y(x) = \frac{x}{2} \sin x - \frac{\pi}{4} \sin x.$$

It is easy to see that this solution satisfies the orthogonality condition to the eigenfunction $\sin x$.

Note that according to equation (4.2) the generalized Green's function determines the amplitude of forced harmonic oscillations of a homogeneous elastic rod, whose extremities are fixed, under the action of an external force of special form. Namely, the frequency of the external force is the resonance one, i.e. it coincides with the frequency of free oscillations of the rod, under which the amplitude of the free oscillations of the rod is described by the function

$\varphi_0(x)$. However, the spatial distribution of the amplitude of the external force is such that the amplitude of forced oscillations does not increase in time and remains bounded. This is achieved by means of the orthogonality conditions: the right-hand side of the equation, as well as the generalized Green's function and the solution of the nonhomogeneous boundary value problem itself, are orthogonal to the eigenfunction $\varphi_0(x)$.

In the given example, the generalized Green's function $G(x, \xi)$ has the physical meaning of the amplitude y of forced oscillations of an elastic rod with fixed extremities under a periodic external action (of resonance frequency) – the sum of a concentrated force applied to the point ξ and a force distributed along the entire rod with amplitude $\varphi_0(\xi) \varphi_0(x)$.

Note finally that we are considering the ideal case, when the amplitudes of the external action are orthogonal to the amplitudes of the natural oscillations very precisely. In real problems these strict conditions are not met as a rule; this may bring about very large amplitudes of the forced oscillations for frequencies of external action close to resonance frequencies or result in the nonexistence of the solution of the forced oscillation problem considered.

We must also keep in mind that the linear mathematical model considered here (and adequately describing small oscillations) is no longer applicable in the case of large amplitudes and must be replaced by a more complicated non-linear model.

§ 3. Eigenvalue Problems

In §1 we stated the eigenvalue problem, which consists in determining the values of the parameter λ such that there exists a non-trivial solution of the homogeneous equation

$$L[y] + \lambda \varrho(x) y(x) = 0, \tag{4.73}$$

on $[0, l]$ which satisfies the homogeneous boundary conditions

$$\alpha_1 y'(0) + \beta_1 y(0) = 0, \quad \alpha_2 y'(l) + \beta_2 y(l) = 0. \tag{4.74}$$

We shall assume that the function $\varrho(x)$ is positive and continuous on $[0, l]$, while the coefficients involved in the differential operator L satisfy the same requirements as in §2. Then Green's function for the boundary value problem (4.26) exists and is a symmetric function of its arguments.

The eigenvalue problem, i.e. the problem of finding nontrivial solutions of the homogeneous equation (4.73) satisfying homogeneous boundary conditions (4.74), is often called the *Sturm-Liouville* problem, while the nontrivial solutions themselves are the eigenfunctions of this problem.

The eigenfunctions of the Sturm-Liouville problem possess a series of remarkable properties which are widely used not only in solving boundary value problems for ordinary differential equations, but also in order to solve many problems concerning partial differential equations and to find the solution of many other mathematical problems. The easiest way to solve most of these problems is by reducing the boundary value problem (4.73), (4.74) to the Fredholm integral equation of the second kind with symmetric kernel. However, the study of integral equations lies beyond the scope of this book (see [5]). Therefore, we shall state the main properties of eigenvalues and eigenfunctions of the boundary value problem (4.73), (4.74) mostly without proof. At the end of this section, we give a reduction of the boundary value problem under consideration to the Fredholm integral equation of the second kind.

The eigenvalues and eigenfunctions of the boundary value problem (4.73), (4.74) possess the following properties:

1. *There exists an infinite countable set $\{\lambda_n\}$ of eigenvalues and a corresponding infinite sequence $\{y_n(x)\}$ of eigenfunctions.* All the eigenvalues may be numbered in increasing order of their absolute value

$$|\lambda_1| \leq |\lambda_2| \leq \ldots \tag{4.75}$$

2. *To each eigenvalue corresponds, up to a constant factor, only one eigenfunction.* In this sense we say that the rank of the eigenvalues equals one. Indeed, assume that there exist two linearly independent eigenfunctions $y_1(x)$ and $y_2(x)$ (there cannot be more than two since the order of the equation is two). It follows from (4.74) that $\alpha_1 y_i'(0) + \beta_1 y_i(0) = 0$ $(i = 1,2)$. Hence $\Delta(y_1(0), y_2(0)) = 0$ and therefore $y_1(x)$, $y_2(x)$ are linearly dependent.

Remark. In the case of more complicated boundary conditions, the rank of the eigenvalue may equal two (but is no more than two, since the order of the equation is two). Eigenvalues of rank two will appear, in particular, for periodic boundary conditions. As an example, consider the boundary value problem

$$y'' + \lambda y = 0, \quad y(0) = y(2\pi), \quad y'(0) = y'(2\pi). \tag{4.76}$$

As we easily see the eigenvalues $\lambda_n = n^2$ in this problem are of rank two. To each λ_n correspond two linearly independent eigenfunctions $y_{n,(1)}(x) = \sin nx$ and $y_{n,(2)}(x) = \cos nx$.

3. *For the case of the boundary conditions*

$$y(0) = y(l) = 0$$

under the assumption $q(x) \geq 0$, all the eigenvalues of the boundary value problem (4.73), (4.74) are positive: $\lambda_n > 0$.

For the proof, multiply the equation for the eigenfunction $y_n(x)$

$$\frac{d}{dx}\left[p(x)\frac{dy_n}{dx}\right]-q(x)y_n(x)+\lambda_n\varrho(x)y_n(x)=0 \qquad (4.77)$$

by the function $y_n(x)$ and integrate the result over $[0,l]$, obtaining

$$\int_0^l \frac{d}{dx}\left[p(x)\frac{dy_n}{dx}\right]y_n(x)dx - \int_0^l q(x)y_n^2(x)dx + \lambda_n \int_0^l \varrho(x)y_n^2(x)dx=0. \qquad (4.78)$$

Taking the first integral by parts and using the boundary conditions, we finally get

$$\lambda_n \int_0^l \varrho(x)y_n^2(x)dx = \int_0^l p(x)\left[\frac{dy_n}{dx}\right]^2 dx + \int_0^l q(x)y_n^2(x)dx, \qquad (4.79)$$

which proves the statement.

4. *The eigenfunctions $y_n(x)$ constitute an orthogonal system $\{y_n(x)\}$ with weight $\varrho(x)$ on the segment $[0,l]$*

$$\int_0^l y_n(x)y_m(x)\varrho(x)dx=0, \qquad n\neq m. \qquad (4.80)$$

Indeed, since to each eigenvalue corresponds precisely one eigenfunction, it remains for us to consider the case when the eigenfunctions $y_n(x)$ and $y_m(x)$ correspond to various eigenvalues $\lambda_n \neq \lambda_m$.

Writing out the equations corresponding to these eigenfunctions

$$L[y_n]+\lambda_n\varrho(x)y_n=0,$$
$$L[y_m]+\lambda_m\varrho(x)y_m=0 \qquad (4.81)$$

and applying Green's formula, we get

$$\int_0^l \{y_n(x)L[y_m]-y_m(x)L[y_m]\}dx=(\lambda_n-\lambda_m)\int_0^l y_n(x)y_m(x)\varrho(x)dx$$
$$=[p(x)(y_n(x)y_m'(x)-y_m(x)y_n'(x))]|_0^l. \qquad (4.82)$$

After carrying out the substitutions indicated here, we obviously obtain zero, since both eigenfunctions $y_n(x)$ and $y_m(x)$ satisfy the homogeneous boundary conditions (4.74). Since $\lambda_n \neq \lambda_m$, this implies our statement.

5. The V. A. Steklov decomposition theorem. *If the function f(x) is twice continuously differentiable on [0, l] and satisfies the homogeneous boundary conditions (4.74), then it can be expanded into an absolutely and uniformly convergent series on [0, l] with respect to the eigenfunctions $y_n(x)$ of the problem (4.73), (4.74)*

$$f(x) = \sum_{n=1}^{\infty} f_n y_n(x). \tag{4.83}$$

The proof of the Steklov theorem will not be given here. We shall only indicate that the orthogonality condition of the eigenfunctions enables us to determine the coefficients f_n of the decomposition in formula (4.83). Indeed, multiplying both parts of formula (4.83) by $y_m(x)\varrho(x)$ and integrating the result over [0, l] (the term by term integration of the series is possible in view of its uniform convergence), we obtain, using (4.80),

$$f_m = \frac{\int_0^l f(x)y_m(x)\varrho(x)dx}{\int_0^l y_m^2(x)\varrho(x)dx}. \tag{4.84}$$

The expression in the denominator is known as the square of the norm of the eigenfunction and is denoted by

$$\|y_m\|^2 = N_m^2 = \int_0^l y_m^2(x)\varrho(x)dx. \tag{4.85}$$

Since the eigenfunctions are determined up to a constant factor, they are normalized in many cases so as to have $N_m = 1$. In this case the system $\{y_n(x)\}$ is said to be orthonormal.

In conclusion, let us stop briefly on the reduction of the boundary value problem (4.73), (4.74) under consideration to Fredholm integral equations of the second kind with symmetric kernel. According to the assumptions made above, Green's function $G(x, \xi)$ for problem (4.26) exists and is a symmetric function of its arguments. Writing equation (4.73) in the form

$$L[y] = -\lambda\varrho(x)y(x) \tag{4.86}$$

we obtain, on the basis of the results of the previous section,

$$y(x) = -\lambda \int_0^l G(x, \xi)\varrho(\xi)y(\xi)d\xi. \tag{4.87}$$

Relation (4.87) is known as the homogeneous Fredholm integral equation of the second kind. It follows from our considerations that any non-trivial solution of the boundary value problem (4.73), (4.74) also satisfies the integral equation (4.87). On the other hand, if $y(x)$ is a solution of equation (4.87), then Theorem 4.1 implies that $y(x)$ is a solution of the boundary value problem (4.26), where $f(x) = -\lambda \varrho(x)y(x)$, i.e. $y(x)$ is a solution of the problem (4.73), (4.74). Thus we see that the given Sturm-Liouville boundary value problem is equivalent to the integral equation (4.87). All the values of the parameters λ for which a nontrivial solution of equation (4.87) exists are called *eigenvalues* and the corresponding solutions *eigenfunctions of equation* (4.87). If follows from the equivalence established above that the boundary value problem (4.73), (4.74) and the integral equation (4.87) have the same eigenvalues and eigenfunctions.

Since Green's function $G(x, \xi)$ is a symmetric function of its arguments, the kernel $-G(x, \xi)\varrho(\xi)$ of the integral equation 4.87 is non-symmetric in general. However, this equation can easily be reduced to an integral equation with symmetric kernel. Indeed, multiplying (4.87) by $\sqrt{\varrho(x)}$ and denoting $v(x) = \sqrt{\varrho(x)}\,y(x)$, we obtain the integral equation

$$v(x) = \lambda \int_0^l \mathcal{K}(x, \xi)v(\xi)d\xi \qquad (4.88)$$

with symmetric kernel

$$\mathcal{K}(x, \xi) = -G(x, \xi)\sqrt{\varrho(x)}\,\sqrt{\varrho(\xi)}. \qquad (4.89)$$

Obviously, the boundary value problem (4.73), (4.74) and the integral equation (4.88) have the same eigenvalues λ_n; the corresponding eigenfunctions $y_n(x)$ of the boundary value problem (4.73), (4.74) and eigenfunctions $v_n(x)$ of the integral equation (4.88) are related by the formula

$$y_n(x) = \frac{v_n(x)}{\sqrt{\varrho(x)}}. \qquad (4.90)$$

Thus the properties of eigenvalues and eigenfunctions of the Fredholm integral equation of the second kind with symmetric kernel, known from the theory of integral equations, enable us to make the necessary conclusions concerning the eigenvalues and eigenfunctions of the Sturm-Liouville boundary value problems.

Chapter V
Stability Theory

§1. Statement of the Problem

Consider the system of equations

$$\frac{dy}{dt} = F(t, y), \qquad (5.1)$$

in which the unknown is an n-dimensional vector function y with components $y_1, \ldots, y_n$. Choose the initial condition

$$y(0) = y_0. \qquad (5.2)$$

It is known from Chapter 2, §5, that under certain natural smoothness conditions concerning the right-hand side of (5.1), the solution $y = y(t, y_0)$ of problem (5.1), (5.2) is a continuous function of t and y_0 at the point (t, y_0), where t is an arbitrary value in some finite closed interval $[0, T]$. Geometrically, this means (see figure 8, in which the case of a one-dimensional y is presented) that for all $\varepsilon > 0$ there exists a $|\varDelta y_0|$ so small that the integral curve $y = y(t, y_0 + \varDelta y_0)$ is contained in a strip of width 2ε surrounding the integral curve $y = y(t, y_0)$ if $t \in [0, T]$. Thus a small error in the initial conditions does not have an important influence on the character of the process, if the process is considered on a finite time interval $t \in [0, T]$.

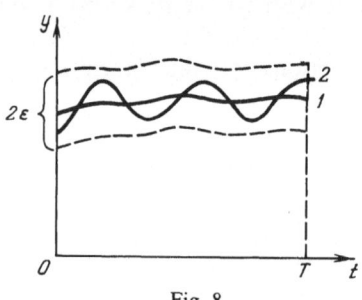

Fig. 8.

1- the integral curve $y = y(t, y_0)$;

2- the integral curve $y = y(t, y_0 + \varDelta y_0)$.

However, it is often necessary to study processes on indefinitely long time intervals; this is expressed mathematically by considering the solution of problem (5.1), (5.2) for $0 \le t < \infty$. We shall assume in advance that the solution of problem (5.1), (5.2) exists on this infinite interval. The following question arises: will the curve $y = y(t, y_0 + \Delta y_0)$ remain in an ε-strip about the curve $y = y(t, y_0)$ for all $t > 0$ when $|\Delta y_0|$ is sufficiently small, or will the curves diverge as t increases?

Examples show that either of the two possibilities may arise. Consider a simple example. The equation $\dfrac{dy}{dt} = ay - 1$ possesses the solution $y(t, y_0) = 1/a$ for the initial value $y_0 = 1/a$. Now suppose we choose the initial value $y_0 + \Delta y_0$. The solution corresponding to this initial value is given by the formula

$$ y(t, y_0 + \Delta y_0) = \left(y_0 + \Delta y_0 - \frac{1}{a} \right) e^{at} + \frac{1}{a} = \Delta y_0 e^{at} + \frac{1}{a}. $$

Thus we see that $a < 0$ implies $|\Delta y| = \left| y(t, y_0 + \Delta y_0) - \dfrac{1}{a} \right| = |\Delta y_0| e^{at} < \varepsilon$ for all $t \ge 0$ whenever $|\Delta y_0| < \varepsilon$. Yet if $a > 0$, then, for sufficiently large t, the value of $|\Delta y|$ becomes as large as we wish, no matter how small $|\Delta y_0|$.

The integral curve such that all integral curves sufficiently close to it for $t = 0$ remain close to it for all $t \ge 0$ as well is called a *stable integral curve* and the corresponding solution is a *stable solution*. In the converse case, we say that the solution is unstable. In the example under consideration, the solution $y = 1/a$ for $a < 0$ is a stable solution while for $a > 0$ it is unstable. Solutions considered on $[0, \infty]$ are thus divided into two disjoint classes: stable solutions and unstable ones.

The notion of stability of solutions was introduced by A. M. Lyapunov. He also laid the foundations of the study of stability. The ideas of A. M. Lyapunov have continued to be important and are widely used in current research on stability questions.

Now let us give a rigorous definition of the notion of stability. Introduce the notation $\|y\| = \sqrt{y_1^2 + \ldots + y_n^2}$, where $y_i (i = 1, \ldots, n)$ are the coordinates of the vector function y.

Definition. The solution $y = y(t, y_0)$ of the problem (5.1), (5.2) is said to be *stable in the sense of Lyapunov*, or *Lyapunov stable*, if $\forall \varepsilon > 0$, $\exists \delta(\varepsilon)$ such that $\forall t > 0$, $\|\Delta y_0\| < \delta(\varepsilon)$ implies the inequality

$$ \|y(t, y_0 + \Delta y_0) - y(t, y_0)\| < \varepsilon. \tag{5.3} $$

(In this definition and later, we will be concerned with stability with respect to initial data. Similarly we could have introduced the notion of stability with respect to parameters appearing in the right-hand side of the equation.)

Among stable solutions we may encounter solutions which possess the following property: all the solutions which are close to the given one at the initial moment of time not only do not diverge from it in the course of time, but even become infinitely close to the given one. Hence we shall give another

Definition. The solution $y=y(t,y_0)$ of the problem (5.1), (5.2) is said to be *asymptotically stable* if 1) it is stable and 2) there exists a sufficiently small number $\delta_0>0$ such that $\|\Delta y_0\|<\delta_0$ implies

$$\lim_{t\to\infty} (y(t,y_0+\Delta y_0)-y(t,y_0))=0. \tag{5.4}$$

In order to find out if the solution $y(t,y_0)$ is stable, we can reduce this question to studying the stability of the trivial (i.e. identically zero) solution of a certain other system related to (5.1). Indeed, let us pass from the unknown y to the new unknown x according to the formula $x=y-y(t,y_0)$. Then the system (5.1) acquires the form

$$\frac{dx}{dt}=f(t,x), \tag{5.5}$$

where $f(t,x)=F(t,x+y(t,y_0))-\dfrac{d}{dt}\,y(t,y_0)$.

To the solution $y(t,y_0)$ in the old variables now corresponds the solution $x\equiv0$ of the system (5.5). Denote

$$x_0=y(0,y_0+\Delta y_0)-y(0,y_0)=\Delta y_0,\quad x(t,x_0)=y(t,y_0+\Delta y_0)-y(t,y_0).$$

Then, in the variables t,x, the definitions of stability and asymptotic stability acquire the following form.

Definition. The trivial solution of the system (5.5) is said to be *Lyapunov stable*, if for $\forall\varepsilon>0$, $\exists\delta(\varepsilon)$ such that $\forall t>0$, $\|x_0\|<\delta(\varepsilon)$ implies the inequality

$$\|x(t,x_0)\|<\varepsilon. \tag{5.6}$$

Definition. The trivial solution of the system 5.5 is said to be *asymptotically stable* if 1) it is stable and 2) $\exists\delta_0>0$ such that $\|x_0\|<\delta_0$ implies

$$\lim_{t\to\infty} x(t,x_0)=0. \tag{5.7}$$

Remark. Sometimes we omit the dependence on x_0 in the expression $x(t,x_0)$ and simply write $x(t)$; then x_0 may be written as $x(0)$; in this case stability means that

$$\|x(t)\|<\varepsilon \tag{5.8}$$

when

$$\|x(0)\| < \delta(\varepsilon), \qquad (5.9)$$

while asymptotic stability means that we also have

$$\lim_{t \to \infty} x(t) = 0, \qquad (5.10)$$

if $\|x(0)\| < \delta_0$.

In the sequel, we shall be concerned with the study of the stability of trivial solutions. The stability of a trivial solution has a convenient geometric interpretation not only in the $(n+1)$-dimensional space of the variables t, x but also in the n-dimensional phase space of the variables x (the notion of phase space was introduced in Chapter 1). We can present a picture of the situation here for the case $n = 2$ (Figure 9). The trivial solution in the phase space is represented

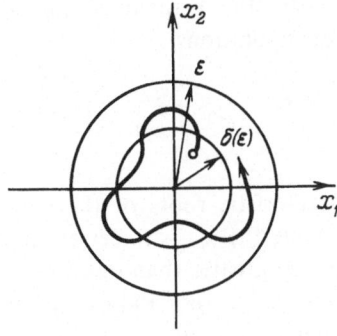

Fig. 9

by a point – the origin. Inequality (5.8) in the two-dimensional case means that the phase trajectory for $t > 0$ is contained within a circle of radius ε with centre at the origin, while inequality (5.9) means that the initial point of the trajectory is located within a circle of radius $\delta(\varepsilon)$, i.e. the trajectory beginning in the δ-neighbourhood of the origin will not leave the ε-neighbourhood of the origin for all $t > 0$; in the case of asymptotic stability, the trajectory will tend to the origin as $t \to \infty$.

Remark. Instead of speaking of the stability of a trivial solution, one often speaks of the stability of the point $(0, \dots, 0)$ of phase space.

Consider the system of two linear homogeneous equations with constant coefficients

$$\frac{dx}{dt} = Ax, \qquad (5.11)$$

where A is a constant (2×2)-matrix

$$A = \begin{pmatrix} a_{11} & a_{12} \\ a_{21} & a_{22} \end{pmatrix}.$$

The system (5.11) has the trivial solution $x = 0$. Let us study its stability. Since a solution of the system (5.11) satisfying arbitrary intial conditions can be written out explicitly, one can readily find out if the trivial solution is stable or unstable. In the general case this cannot be done. However, results which we shall obtain for (5.11) will indicate the direction of study in the general case.

First let us obtain some auxiliary inequalities.

According to the results of §§6 and 7 in Chapter 3, the solution $x(t, x_0)$ of system (5.11) with initial value x_0 can be represented in the form

$$x(t, x_0) = \mathscr{K}(t, 0) x_0, \tag{5.12}$$

where $\mathscr{K}(t, 0) = W(t) W^{-1}(0)$; $W(t)$ is the fundamental matrix, whose columns are two linearly independent solutions

$$x_{(1)} = \begin{pmatrix} \alpha_{(1)1} \\ \alpha_{(1)2} \end{pmatrix} e^{\lambda_1 t}, \qquad x_{(2)} = \begin{pmatrix} \alpha_{(2)1} \\ \alpha_{(2)2} \end{pmatrix} e^{\lambda_2 t}. \tag{5.13}$$

Here λ_1, λ_2 are the characteristic roots of the matrix A (the roots of the characteristic equation), while the $\alpha_{(i),j}$ are certain numbers when $\lambda_1 \neq \lambda_2$ and polynomials in t of degree no greater than one when $\lambda_1 = \lambda_2$.

Suppose $\lambda_j = p_j + i q_j$. Then $|e^{\lambda_j t}| = e^{p_j t}$ and we have the following estimate for the elements $\mathscr{K}_{ij}$ of the matrix $\mathscr{K}(t, 0)$:

$$|\mathscr{K}_{ij}| < (C_1 + C_2 t) e^{pt}, \tag{5.14}$$

where $p = \max \{p_1, p_2\}$; C_1, C_2 are certain constants and $C_2 = 0$ in the case $\lambda_1 \neq \lambda_2$. The inequality (5.14) may also be written (taking into consideration the fact that for any $\gamma > 0$ we have $te^{-\gamma t} \leq \dfrac{1}{\gamma} e^{-1} = C_3$) in the following form

$$|\mathscr{K}_{ij}| < (C_1 + C_2 t) e^{-\gamma t} e^{(p+\gamma)t} \leq (C_1 + C_2 C_3) e^{(p+\gamma)t},$$

i.e.

$$|\mathscr{K}_{ij}| < C e^{(p+\gamma)t}. \tag{5.15}$$

Let us derive one more auxiliary inequality. Suppose $y = A(t)x$, where y, x are columns and $A(t)$ is a square matrix. Suppose the elements of $A(t)$ satisfy the inequality $|a_{ij}(t)| < a(t)$. Then

$$\|y\| \leq 2a(t) \|x\|. \tag{5.16}$$

Indeed, for the two-dimensional case, we have $y_1 = a_{11}x_1 + a_{12}x_2$ and therefore

$$y_1^2 = a_{11}^2 x_1^2 + a_{12}^2 x_2^2 + 2a_{11}a_{12}x_1 x_2$$
$$\leq a_{11}^2 x_1^2 + a_{12}^2 x_2^2 + |a_{11}||a_{12}|(x_1^2 + x_2^2) \leq 2a^2(x_1^2 + x_2^2).$$

Taking into consideration a similar estimate for y_2, we obtain $y_1^2 + y_2^2 \leq 4a^2(x_1^2 + x_2^2)$, which is equivalent to (5.16).

Applying inequality (5.16) to (5.12) and taking the right-hand side of (5.14) or (5.15) for $a(t)$, we obtain (the coefficient 2 may be included in the constants C_i or C).

$$\|x(t, x_0)\| \leq (C_1 + C_2 t)e^{pt}\|x_0\|, \tag{5.17}$$

$$\|x(t, x_0)\| \leq Ce^{(p+\gamma)t}\|x_0\|. \tag{5.18}$$

Now let us turn to the actual study of the trivial solution of system (5.11). We consider several cases.

1. Suppose $p_1 = \operatorname{Re} \lambda_1 < 0$, $p_2 = \operatorname{Re} \lambda_2 < 0$. Then $p = \max\{p_1, p_2\} < 0$ and therefore we have $p + \gamma < 0$ for a sufficiently small γ. Then for all $\varepsilon > 0$ we obtain from (5.18) the inequality $\|x(t, x_0)\| \leq C\|x_0\| < \varepsilon$ when $\|x_0\| < \varepsilon/C = \delta(\varepsilon)$, so that the trivial solution of the system (5.11) is stable. We also see from (5.18) that $x(t, x_0) \to 0$ as $t \to \infty$, and therefore the solution is asymptotically stable.

2. Suppose at least one of the numbers p_1, p_2 is positive, say p_1. In this case, as can be easily seen, the trivial solution of the system (5.11) is not stable. Assume the converse, i.e. suppose that for $\forall \varepsilon > 0$ $\exists \delta(\varepsilon)$ such that $\|x_0\| < \delta(\varepsilon)$ implies $\|x\| < \varepsilon$. Further the argument is different in the cases when the λ_i are real or complex conjugate. For real λ_i it is sufficient to consider the solution $x = Cx_{(1)}$. For sufficiently small $|C|$, we obviously have $\|x_0\| < \delta(\varepsilon)$, but the inequality $\|x\| < \varepsilon$ cannot hold for all $t > 0$ since $e^{p_1 t} \to \infty$ as $t \to \infty$. In the case when the λ_i are complex, we have $p_1 = p_2 = p$, $\lambda_1 = p + iq$ and we may consider the solution $x = C \operatorname{Re} x_{(1)}$. For a sufficiently small $|C|$, we still have $\|x_0\| < \delta(\varepsilon)$. Choose one of the components of this solution which is not identically zero, say $x_1 = C\beta_1(t)e^{pt}$, where $\beta_1(t)$ is some well determined linear combination of $\cos qt$ and $\sin qt$. Obviously x_1 is not bounded as $t \to \infty$ and therefore the inequality $\|x\| < \varepsilon$ cannot hold for all $t > 0$.

3. It remains for us to consider the case when $p_1 = 0, p_2 < 0$ or $p_1 = p_2 = 0$. The latter means that λ_1 and λ_2 are either purely imaginary or equal to zero and multiple. In the case $p_1 = 0, p_2 < 0$ or the case of purely imaginary λ_j in formula (5.17), we have $p = 0, C_2 = 0$ so that $\|x(t, x_0)\| \leq C_1\|x_0\|$ and the solution is stable for the same reasons as in 1. However, asymptotic stability will not occur here since we can obviously find solutions which do not tend to zero as $t \to \infty$. In the case $p_1 = 0, p_2 < 0$, such a solution will be, say, a solution of the form $x = Cx_{(1)}$ for which the value $\|x_0\|$ is as small as we wish for sufficiently small $|C|$ but $\|x\| = \text{const} \not\to 0$ as $t \to \infty$. For purely imaginary λ_i, obviously $x = C \operatorname{Re} x_{(1)} \not\to 0$ for $t \to \infty$.

In the case $\lambda_1 = \lambda_2 = 0$, we have $e^{\lambda_1 t} = e^{\lambda_2 t} = 1$, while the $\alpha_{(j)i}$ are linear functions of t and therefore the inequality $\|x(t, x_0)\| < \varepsilon$ cannot hold for all $t > 0$, except in the case when the $\alpha_{(j)i}$ degenerate into polynomials of degree zero. But this case takes place only when $a_{ik} = 0$, the solution will then be of the form $x_1 = x_{01}$, $x_2 = x_{02}$ and is obviously stable, but not asymptotically stable.

Remark. It is easy to see that the analysis carried out above will remain valid to a great extent in the n-dimensional case. Thus the estimate (5.15) for $\mathscr{K}_{ij}$ and inequality (5.16) (in which instead of the coefficient 2 we shall have an expression depending on n) both remain valid. And so will inequality (5.18). It is also clear that the arguments carried out in 1, 2 can be carried over without any changes to the n-dimensional case.

Summarizing our results, we see that *the trivial solution of a homogeneous system of linear equations is stable and asymptotically stable if* $\mathrm{Re}\,\lambda_i < 0$ *for all i and non-stable if* $\mathrm{Re}\,\lambda_i > 0$ *for at least one i.*

In the case of characteristic roots with vanishing real parts, the situation becomes more complicated. Further analysis shows that if the characteristic roots with vanishing real parts are not multiple roots, then, under the condition that the other λ satisfy the inequality $\mathrm{Re}\,\lambda < 0$, the solution will be stable but not asymptotically. In the case when there are multiple roots among the λ with vanishing real parts, stability will not occur in general, even if the other λ satisfy the inequality $\mathrm{Re}\,\lambda < 0$.

§2. Study of Stability in the First Approximation

Consider a system of the general form (5.5) in the so-called autonomous case, i.e. when f does not explicitly depend on t. Let us write the system in coordinate form

$$\frac{dx_i}{dt} = f_i(x_1, \ldots, x_n) \quad (i = 1, \ldots, n). \tag{5.19}$$

Since the notion of stability of the trivial solution is related to a small neighbourhood of the origin in the phase space, it is natural to expect that the behaviour of the solution of the system (5.19) will be determined by the principal terms of the decomposition of f with respect to x in a neighbourhood of $x = 0$. Since $f_i(0, \ldots, 0) = 0$ ($x \equiv 0$ being a solution of system (5.19)), the principal terms will be the linear terms of f with respect to x, or, as they are otherwise called, the terms of the first approximation.

According to Taylor's formula, if we take into consideration the fact that $f_i(0, \ldots, 0) = 0$, we get

$$f_i(x_1, \ldots, x_n) = \sum_{k=1}^{n} a_{ik} x_k + R_{(2)i}, \tag{5.20}$$

where

$$a_{ik} = \frac{\partial f_i}{\partial x_k}(0, \dots, 0),$$

$$R_{(2)i} = \frac{1}{2} \sum_{j,l=1}^{n} \frac{\partial^2 f_i}{\partial x_j \partial x_l}(\theta x_1, \dots, \theta x_n) x_j x_l. \tag{5.21}$$

If we omit $R_{(2),1}$ in (5.20), then, instead of (5.19), we obtain a linear system of the form

$$\frac{dx_i}{dt} = \sum_{k=1}^{n} a_{ik} x_k, \tag{5.22}$$

which we shall call the first approximation system for the system (5.19). The behaviour of the system (5.22) with respect to stability or nonstability of the trivial solution is determined by the properties of the roots λ of the characteristic equation, as we have learned at the end of the previous section. We may expect that the same requirements concerning λ will guarantee not only the stability or the nonstability of the trivial solution of the system (5.22), but also that of the trivial solution of the given system (5.19). As we shall see further, this prediction is justified. However, before we state the corresponding theorem, let us prove a lemma containing some auxiliary inequalities which shall be used to prove the theorem. Further, following a common practice, all the positive constants, whose actual values are unimportant, will be denoted by the same letter C (thus in 5.17) we could have written $\|x(t, x_0)\| \leq (C + Ct)e^{pt}\|x_0\|$, etc).

Lemma 5.1. *We have the following statements:*

1°. Suppose y is a vector with components $y_i = \sum_{k=1}^{n} a_{ik}(t)x_k$, where $|a_{ik}(t)| < a(t)$.

Then

$$\|y\| < Ca(t)\|x\|.$$

2°. Suppose y is a vector with components $y_i = \sum_{j,l=1}^{n} a_{ijl}(t)x_j x_l$, where $|a_{ijl}(t)| < a(t)$. Then

$$\|y\| < Ca(t)\|x\|^2.$$

3°. For any pair of vectors x and y, we have the inequality

$$\|x+y\| \leq C(\|x\| + \|y\|).$$

4°. For any vector y, we have the inequality

$$\left\| \int_0^t y\,d\xi \right\| \leq C \int_0^t \|y\|\,d\xi.$$

$5°$. *For the matrizant $\mathcal{K}(t, \xi)$ of the linear system (5.22), we have the inequality*

$$|\mathcal{K}_{ij}(t, \xi)| = |\mathcal{K}_{ij}(t - \xi, 0)| < Ce^{(p+\gamma)(t-\xi)}, \tag{5.23}$$

where $p = \max_{i=1,\ldots,n} (\operatorname{Re} \lambda_i)$ and γ is a positive constant.

Proof. $1°$. The statement was proved in the two-dimensional case (see 5.16). As was noted in the previous section, the proof is similar for an arbitrary n.

$2°$. The statement is proved by means of similar considerations and the proof is omitted.

$3°$. We have

$$(x+y)_i = x_i + y_i \Rightarrow [(x+y)_i]^2 = x_i^2 + y_i^2 + 2x_i y_i$$

$$\leq \|x\|^2 + \|y\|^2 + 2\|x\| \|y\| = (\|x\| + \|y\|)^2 \Rightarrow \|x+y\|^2$$

$$\leq n(\|x\| + \|y\|)^2 \Rightarrow \|x+y\| \leq C(\|x\| + \|y\|) \quad (C = \sqrt{n}).$$

$4°$. We have

$$\left(\int_0^t y d\xi\right)_i = \int_0^t y_i d\xi \Rightarrow \left[\left(\int_0^t y d\xi\right)_i\right]^2$$

$$\leq \left(\int_0^t \|y\| d\xi\right)^2 \Rightarrow \left\|\int_0^t y d\xi\right\| \leq C \int_0^t \|y\| d\xi \quad (C = \sqrt{n}).$$

Remark. The inequalities written above easily follow from linear algebra theorems.

$5°$. For $\mathcal{K}(t, \xi)$ we have inequality (5.15):

$$|\mathcal{K}_{ij}(t, 0)| < Ce^{(p+\gamma)t} \tag{5.24}$$

(see the remark at the end of §1). Let us show that $\mathcal{K}(t, \xi) = \mathcal{K}(t - \xi, 0)$. Indeed, $\mathcal{K}(t, \xi)$ satisfies the equation $\dfrac{d}{dt} \mathcal{K}(t, \xi) = AK(t, \xi)$ (A is the matrix with elements a_{ik}) and the initial condition $\mathcal{K}|_{t=\xi} = \mathcal{K}(\xi, \xi) = E$. Replacing the independent variable t by $t - \xi$ and taking into consideration the fact that $A = \text{const}$, we see that $\mathcal{K}(t - \xi, 0)$ satisfieds the same equation and the initial condition $\mathcal{K}|_{t-\xi=0} = \mathcal{K}(0, 0) = E$. Hence, by the uniqueness theorem, $\mathcal{K}(t, \xi) \equiv \mathcal{K}(t - \xi, 0)$. Thus (5.24) implies (5.23).

Now let us state a theorem which determines the stability or non-stability of the trival solution of system (5.19) according to the characteristic roots of the first approximation matrix.

Theorem 5.1. *Suppose that in some neighbourhood of the point x_1 $=0,\ldots,x_n=0$ the functions $f_i(x_1,\ldots,x_n)$ $(i=1,\ldots,n)$ are continuous together with their derivatives (up so the second order inclusive). Then, if all the characteristic roots λ_i of the matrix with elements $a_{ik}=\dfrac{\partial f_i}{\partial x_k}$ $(0,\ldots,0)$ satisfy the condition $\operatorname{Re}\lambda_i<0$, the trivial solution of this system (5.19) is stable and even asymptotically stable. In the case when $\operatorname{Re}\lambda_i>0$ for at least one i, the trivial solution of the system (5.19) is unstable.*

We shall limit ourselves to proving the first statement of the theorem concerning stability and asymptotic stability.

Using the representation (5.20) for the right-hand sides of (5.19) and viewing (5.19) as the system (3.71) in which $f=R_{(2)}$, let us pass from the differential equation (5.19) with initial condition $x(0)=x_0$ to the equivalent integral equation (see (3.87))

$$x=\mathscr{K}(t,0)x_0+\int_0^t \mathscr{K}(t,\xi)R_{(2)}(\xi)d\xi. \tag{5.25}$$

Now consider some neighbourhood Ω of the point $x=0$ (in phase space) given by the inequality $\|x\| < K$, where K is some constant. According to Lemma 5.1 we have the estimate $\|R_{(2)}\| < C\|x\|^2$ in the domain Ω, and we can pass from (5.25) to the inequality

$$\|x\| \le Ce^{-\alpha t}\|x_0\|+C\int_0^t e^{-\alpha(t-\xi)}\|x\|^2 d\xi, \tag{5.26}$$

where $-\alpha=p+\gamma$. Since $p<0$ in the case considered, we also have $\alpha>0$ for sufficiently small γ.

Now consider the auxiliary scalar problem

$$\frac{dz}{dt}=-\alpha z+Cz^2, \quad z(0)=z_0>C\|x_0\|. \tag{5.27}$$

This equation possesses an elementary integral (as an equation with separable variables or as a Bernoulli equation) and its solution is of the form

$$z=\frac{\alpha z_0}{Cz_0+(\alpha-Cz_0)e^{\alpha t}}.$$

This solution, as can be easily seen, has the following properties:
1) $z>0$ for $t\ge0$ if z_0 is sufficiently small: $z_0<\alpha/C$;
2) for all $\varepsilon>0$, $z<\dfrac{\alpha z_0 e^{-\alpha t}}{\alpha-Cz_0}<\varepsilon$ if $z_0<\dfrac{\varepsilon\alpha}{\alpha+\varepsilon C}=\delta(\varepsilon)$;
3) $z(t)\to0$ as $t\to\infty$.

Now let us write out an integral equation of the same type as (5.25) for $z(t)$:

$$z = z_0 e^{-\alpha t} + C \int_0^t e^{-\alpha(t-\xi)} z^2 d\xi \qquad (5.28)$$

and compare $\|x\|$ with z. We claim that, for $t \geq 0$, we have the inequality

$$\|x\| < z. \qquad (5.29)$$

Indeed, for $t = 0$ the inequality (5.29) is valid, since by putting $t = 0$ in (5.26) we obtain $\|x_0\| < z_0$. Assume that for some value $t = t_1$ the inequality (5.29) ceases to be valid and we have the relation $\|x(t_1)\| = z(t_1)$. It follows from property 2) of the function $z(t)$ that, for a sufficiently small z_0, we have $|z| \leq K$ for $t \geq 0$. If $0 \leq t \leq t_1$, (5.29) holds and therefore $\|x\| \leq K$ (i.e. $x \in \Omega$). Thus, for $0 \leq t \leq t_1$, we have the inequality (5.26) and, for $t = t_1$,

$$z(t_1) = z_0 e^{-\alpha t_1} + C \int_0^{t_1} e^{-\alpha(t_1-\xi)} z^2 d\xi = \|x(t_1)\| \leq C \|x_0\| e^{-\alpha t_1}$$

$$+ C \int_0^{t_1} e^{-\alpha(t_1-\xi)} \|x\|^2 d\xi < z_0 e^{-\alpha t_1} + C \int_0^{t_1} e^{-\alpha(t_1-\xi)} z^2 d\xi.$$

Comparing the first and last terms of this chain of inequalities, we obtain a contradiction in the form $1 < 1$; this proves the validity of (5.29) for $t > 0$.

And now, using inequality (5.29), it is easy to obtain the theorem. Indeed, suppose we are given any $\varepsilon > 0$. Put $\bar{\delta}(\varepsilon) = \dfrac{\delta(\varepsilon)}{2C}$, where $\delta(\varepsilon)$ is the number which appears in property 2) of the function $z(t)$. Suppose $\|x_0\| < \bar{\delta}(\varepsilon)$. Take $z_0 = \frac{1}{2}\delta(\varepsilon) < \delta(\varepsilon)$. We then have $z(t) < \varepsilon$. But since $C\|x_0\| < C\bar{\delta}(\varepsilon) = \dfrac{\delta(\varepsilon)}{2} = z_0$, the relation (5.29) is valid, so that $\|x(t)\| < z(t) < \varepsilon$ for $\|x_0\| < \bar{\delta}(\varepsilon)$, i.e. the trivial solution of the system (5.19) is stable. The same inequality (5.29), together with property 3) of the function $z(t)$, guarantees the asymptotic stability of the trivial solution of system (5.19).

Remarks. 1. We have proved the asymptotic stability theorem. The following statement on asymptotic non-stability also holds: if $\mathrm{Re}\, \lambda_i > 0$ for at least one i, then the trivial solution of system (5.19) is unstable.

2. The statements on asymptotic stability and on non-stability remain valid in the case when f depends explicitly on t, as long as the representation $f_i(x, t) = \sum_{k=1}^n a_{ik} x_k + R$, where $a_{ik} = \text{const}$ and $\|R\| < C \|x\|^{1+\alpha}$ ($\alpha > 0$ being arbitrary), is valid.

3. If among the λ there are numbers satisfying Re $\lambda = 0$, then, even if the trivial solution of the system of the first approximation (5.22) is stable, the trivial solution of the system (5.19) may be either stable or unstable, depending on R. In this case, which is usually called the *critical case*, it is impossible to reach a conclusion concerning the stability or non-stability of the trivial solution of the system (5.19) if we only know the characteristic roots of the first approximation matrix. In the critical cases, the study of stability according to the first approximation becomes meaningless and one must use the properties of the subsequent terms of the decomposition (5.20), or other methods.

§ 3. The Method of Lyapunov Functions

The method for studying stability developed in the previous paragraph, despite its naturality, does not always give an answer to the question under consideration.

A. M. Lyapunov also proposed a different method. In this method, the given system of equations is assigned a certain function in the variables $x_1, \ldots, x_n$, called a Lyapunov function, and its properties are used to draw conclusions concerning the stability of the solution.

Let us illustrate the idea of the method for the following simple example

$$\frac{dx_1}{dt} = -x_1 + x_2 = f_1, \qquad \frac{dx_2}{dt} = -2x_2 = f_2. \qquad (5.30)$$

We know from the above that the trivial solution of this system is stable, since $\lambda_1 = -1 < 0$, $\lambda_2 = -2 < 0$. However, in order to check the stability of the trivial solution, it is possible to argue in a different way. Consider the function $V(x_1, x_2) = 2x_1^2 + x_2^2$. This function is positive everywhere except at the point $x_1 = 0$, $x_2 = 0$, where it vanishes. In the space of the variables x_1, x_2, V, the equation $V = 2x_1^2 + x_2^2$ determines a paraboloid with summit at the origin. The level lines of this surface on the (x_1, x_2)-plane are ellipses. Let us choose an arbitrarily small ε. In the (x_1, x_2)-plane, consider the circle ω_ε of radius ε. Choose one of the level lines, an ellipse entirely contained in the circle ω_ε. Construct another circle ω_δ, entirely located within the ellipse (Fig. 10). Suppose the initial point $A(x_{1,0}, x_{2,0})$ lies within ω_δ.

Consider the function in two variables $W(x_1, x_2) = (\text{grad } V, f)$. It is easy to see that if we substitute the solution $x_1(t)$, $x_2(t)$ of system (5.30) for the variables x_1, x_2, then the function of t thus obtained will be the complete derivative $\dfrac{dV}{dt}$ of $V(x_1(t), x_2(t))$ along the trajectories of the solution of the system (5.30). If this derivative is non-positive along any trajectory which originates in ω_δ, then this will mean that such a trajectory cannot leave ω_ε, since otherwise, between $t = 0$

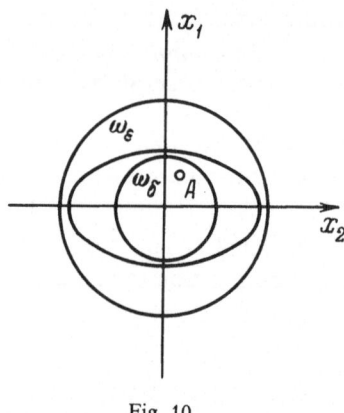

Fig. 10

and the value $t = t_1$ for which it reaches the boundary of ω_ε, we will be able to find a value $t = t^*$ such that $\dfrac{dV}{dt} > 0$, since $V(x_1(t_1), x_2(t_1)) > V(x_{1,0}, x_{2,0})$. The fact that any trajectory originating in ω_δ never leaves the circle ω_ε for any $t > 0$ means that the trivial solution is stable.

Thus we must check the sign of $\dfrac{dV}{dt}$ along the trajectory. In order to do this, we must know the trajectory itself. Although in the given example this can be done, the method must be applicable to systems of general form, for which $x_1(t)$, $x_2(t)$ cannot be written out explicitly in order to check the necessary inequality. Therefore, we shall require the function $W(x_1, x_2)$ to be non-positive (as a function of two independent variables x_1, x_2) at least in some neighbourhood of $(0,0)$. This condition may be checked readily from the right-hand side of the system without knowing the solution. This will be the case in our example, since

$$W(x_1, x_2) = -2\,[(x_1 - x_2)^2 + x_1^2 + x_2^2] \le 0$$

holds everywhere on the (x_1, x_2)-plane and therefore holds along any trajectory, and the stability of the trivial solution is guaranteed.

The function $V(x_1, x_2)$ which was involved in these arguments is precisely the Lyapunov function for the example under consideration. It is expressed as a quadratic form $2\,x_1^2 + x_2^2$, although in principle instead of $2\,x_1^2 + x_2^2$ we could have taken some other function, as long as it is positive everywhere except at the point $(0,0)$, where it vanishes, the expression $(\operatorname{grad} V, f) \equiv W(x_1, x_2)$ being non-positive. Let us stress once more that, in the arguments above, the fact that the function V is positive, and that the function W, whose value along the trajectories is the complete derivative of V with respect to t along these trajectories, is non-positive, is essential.

Now let us pass to the statement and proof of certain general theorems, which are based on the idea illustrated above. We shall study the trivial solution of the system (5.5):

$$\frac{dx}{dt} = f(t, x). \tag{5.31}$$

All our further constructions will be carried out within a neighbourhood Ω of the origin in the phase space. Suppose, to be specific, that Ω is given by the inequality $\|x\| < K$, where $K > 0$ is some constant.

Definition. The function $V(x_1, \ldots, x_n)$ (or in a shorter form, $V(x)$) is said to be *positive definite* in Ω if $V(x) \geq 0$ in Ω, while $V(x) = 0$ only for $x = 0$.

Lemma 5.2. *Suppose the function $V(x)$ is positive definite and continuous in Ω. Then a) for all x satisfying the inequality $\|x\| \geq \varepsilon_1 > 0$ there exists an $\varepsilon_2 > 0$ such that $V(x) \geq \varepsilon_2$; b) conversely, the inequalities $V(x) \geq \varepsilon_2 > 0$ imply the existence of an $\varepsilon_1 > 0$ such that $\|x\| \geq \varepsilon_1$.*

The proof of both statements will be carried out by *reductio ad absurdum*.

a) Suppose that the inequality $\|x\| \geq \varepsilon_1$ holds and the inequality $V(x) \geq \varepsilon_2$ is not valid for any ε_2. Then $\forall \varepsilon_{(1)2} \exists x_{(1)}$ satisfying the inequality $\|x_{(1)}\| \geq \varepsilon_1$ and such that $V(x_{(1)}) < \varepsilon_{(1)2}$. Choose a sequence $\varepsilon_{(n)2} \to 0$. To this sequence will correspond a sequence of points $x_{(n)}$ such that $\|x_{(n)}\| \geq \varepsilon_1$, while

$$V(x_{(n)}) < \varepsilon_{(n)2} \to 0. \tag{5.32}$$

We have $\varepsilon_1 \leq \|x_{(n)}\| < K$. Therefore, it is possible to choose a subsequence of the sequence $x_{(n)}$, which, by introducing a new numeration, we shall still denote $x_{(n)}$, and which has the limit $\bar{x}$; then we have

$$\varepsilon_1 \leq \|\bar{x}\| \leq K. \tag{5.33}$$

In view of the continuity of $V(x)$, we have $\lim_{n \to \infty} V(x_{(n)}) = V(\bar{x})$. But the inequality (5.32) implies $\lim_{n \to \infty} V(x_{(n)}) = 0$. Thus $V(\bar{x}) = 0 \Rightarrow \bar{x} = 0$, which contradicts (5.33).

b) Suppose that the inequality $V(x) \geq \varepsilon_2$ holds and the inequality $\|x\| \geq \varepsilon_1$ does not hold for any ε_1. Then $\forall \varepsilon_{(1)1} \exists x_{(1)}$ satisfying the inequality $\|x_{(1)}\| < \varepsilon_{(1)}$, while $V(x_{(1)}) \geq \varepsilon_2$. Choose a sequence $\varepsilon_{(n),1} \to 0$. To this sequence will correspond a sequence of points $x_{(n)}$ such that $V(x_{(n)}) \geq \varepsilon_2$, while $\|x_{(n)}\| \leq \varepsilon_{(n),1} \to 0$. But the last relation means that $x_{(n)} \to 0$. Hence, by the continuity of $V(x)$, we have $\lim_{n \to \infty} V(x_{(n)}) = V(0) = 0$. On the other hand, $\lim_{n \to \infty} V(x_{(n)}) \geq \varepsilon_2$, i.e. $0 \geq \varepsilon_2$ (contradiction).

Theorem 5.2 *(stability theorem). Suppose there exists a function $V(x)$, continuous together with its partial derivatives of the first order and positive definite in Ω, such that the function $W(x,t) \equiv (\operatorname{grad} V, f(t,x))$ satisfies the inequality*

$$W(x,t) \leq 0 \quad for \quad t > 0, \quad x \in \Omega. \tag{5.34}$$

Then the trivial solution of the system (5.31) is stable.

Proof. Choose an arbitrary $\varepsilon > 0$. Lemma 5.2 guarantees (putting $\varepsilon_1 = \varepsilon$) the existence of an $\varepsilon_2(\varepsilon)$ such that $\|x\| > \varepsilon$ implies $V(x) \geq \varepsilon_2$. Further, by the continuity of $V(x)$ (for $x=0$), we can find a $\delta_1(\varepsilon_2) = \delta(\varepsilon)$ such that $\|x\| < \delta$ implies $V(x) \leq \varepsilon_2/2$.

Consider the initial point $x(0) = x_0$ in the δ-ball ω_δ of the phase space of the variables x (Fig. 11 pictures the situation in the two-dimensional case), i.e. suppose that $\|x(0)\| < \delta$. Then $V(x(0)) \leq \varepsilon_2/2$. We must prove that the trajectory remains within the ε-ball ω_ε for all $t > 0$ (see (5.8)).

Assume the converse, i.e. suppose the trajectory leaves ω_ε for some $t = t_1$, remaining in Ω. Then $V(x(t_1)) \geq \varepsilon_2$. Thus we have

$$V(x(t_1)) - V(x(0)) \geq \varepsilon_2 - \frac{\varepsilon_2}{2} = \frac{\varepsilon_2}{2} > 0. \tag{5.35}$$

On the other hand,

$$V(x(t_1)) - V(x(0)) = \frac{dV}{dt}\Big|_{t=t^*} t_1 = \left(\operatorname{grad} V, \frac{dx}{dt}\right)_{t=t^*} t_1$$

$$= (\operatorname{grad} V, f(t^*, x(t^*))) t_1 = W(x(t^*), t^*) t_1 \leq 0 \quad (0 \leq t^* \leq t_1), \tag{5.36}$$

because (5.34) holds everywhere in Ω and hence along the trajectory as well. The inequalities (5.35) and (5.36) are in contradiction, which proves the theorem.

Theorem 5.3 *(asymptotic stability theorem). Suppose that in addition to the assumptions of Theorem 5.2 for $t \geq 0$, $x \in \Omega$, we have the inequality $W(x,t) \leq -\bar{W}(x)$, where $\bar{W}(x)$ is a positive definite function in Ω. Then the trivial solution of the system (5.31) is asymptotically stable.*

Proof. The stability of the trivial solution follows from the previous theorem. Let us check that

$$\lim_{t \to \infty} x(t) = 0. \tag{5.37}$$

Using the expression for the complete derivative of V along the trajectory, we obtain $\dfrac{dV}{dt} = W(x(t), t) \leq -\bar{W}(x(t)) \leq 0$, i.e. $V(x(t))$ does not increase monotonically when t increases. Therefore the limit $\lim_{t \to \infty} V(x(t)) = \alpha \geq 0$ exists. Let us

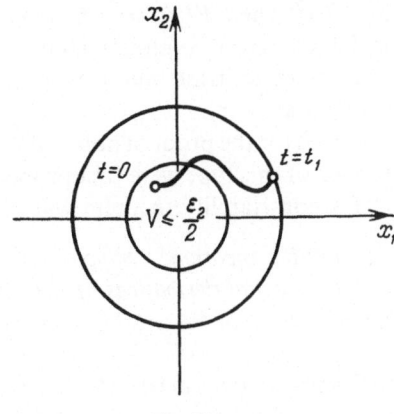

Fig. 11

prove that $\alpha = 0$. Assume that $\alpha > 0$. Since $V(x(t)) \geq 0$ (V tends to an upper limit), by putting $\varepsilon_2 = \alpha$ in part b) of Lemma 5.2, we obtain $\|x(t)\| \geq \varepsilon_1$ and then, using the same lemma (part a)) we get $\bar{W}(x(t)) \geq \beta > 0$, i.e. $-\bar{W}(x(t)) \leq -\beta < 0$. Therefore

$$V(x(t)) - V(x(0)) = \frac{dV}{dt}\bigg|_{t=t^*} \qquad t \leq -\bar{W}(x(t^*))t \leq -\beta t.$$

Thus we see that $V(x(t)) \to -\infty$ as $t \to \infty$. On the other hand, since (by stability) we have $x(t) \in \Omega$, it follows that $V(x(t)) \geq 0$. The contradiction obtained shows that $\alpha = 0$. Thus

$$\lim_{t \to \infty} V(x(t)) = 0. \tag{5.38}$$

Let us prove that this implies (5.37).

Assume the converse. Then there exists an $\varepsilon_1 > 0$ and a sequence $t_n \to \infty$ satisfying $\|x(t_n)\| \geq \varepsilon_1$. But then, according to Lemma 5.2, $V(x(t_n)) \geq \varepsilon_2$, which contradicts (5.38). The contradiction proves (5.37), concluding the proof of the theorem.

Examples. In the example (5.30) considered above, we obtain not only stability, but asymptotic stability as well, since $-W(x_1, x_2)$ does not depend on t and is a positive definite function. This example, however, cannot demonstrate the importance of theorems (5.2) and (5.3) proved above, since the arguments of the previous section may be used in it – asymptotic stability follows from the fact that λ_1 and λ_2 are negative.

Let us give an example of a system in which Theorem 5.1 concerning the first answer approximation cannot be applied, while the Lyapunov function gives the

$$\frac{dx_1}{dt} = 2x_2 - x_1^3 \sin^2 t, \qquad \frac{dx_2}{dt} = -3x_1 - x_2^5.$$

Choose $V(x) = 3 x_1^2 + 2 x_2^2$. Then $W(x, t) = -6 x_1^4 \sin^2 t - 4 x_2^6 \leq 0$. Therefore, according to Theorem 5.2, the trivial solution is stable. Theorem of 5.1 gives no answer here, since the characteritistic numbers of the matrix of the first approximation are purely imaginary.

Similar ideas may be used for the proof of non-stability of the trivial solution of system (5.31). Theorems of this type were first proposed by N. G. Chetayev. The simplest version of a non-stability theorem will be proposed here.

Theorem 5.4 *(non-stability theorem). Suppose the function $V(x)$ is continuous, together with all its partial derivatives of the first order in Ω and:*

a) In any δ-neighbourhood ω_δ of the origin, there exists a point x such that $V(x) > 0$.

b) If ω_ε is an ε-neighbourhood of the origin, then for $\forall \alpha > 0 \; \exists \beta > 0$ such that the condition $x \in \omega_\varepsilon$, $V(x) \geq \alpha$ implies the inequality $W(x, t) = (\operatorname{grad} V, f(t, x)) \geq \beta$, which is valid for all $t \geq 0$.

Then the trivial solution of the system (5.31) is unstable.

Before proving the theorem, let us consider an example, namely the system

$$\frac{dx_1}{dt} = x_1 x_2^4, \qquad \frac{dx_2}{dt} = x_2 x_1^2$$

Take $V(x) = x_1 x_2$. Obviously, in any ω_δ there exists a point (for example, in the first quadrant) such that $V = x_1 x_2 > 0$. Suppose the initial point satisfies $x_1^0 x_2^0 = \alpha > 0$. On Fig. 12 the part $\omega_\varepsilon^\alpha$ of the interior of the circle ω_ε bounded by the hyperbola $x_1 x_2 = \alpha$, for which $V \geq \alpha$, is cross-hatched. Obviously in $\omega_\varepsilon^\alpha$ we have the inequality $x_1^2 + x_2^2 \geq 2\alpha$, i.e. $\|x\| \geq \sqrt{2\alpha}$. Now, according to Lemma 5.2, $\exists \gamma > 0$ such that $x_2^4 + x_1^2 \geq \gamma$ and therefore $W(x, t) \geq \alpha \gamma = \beta$. By Theorem 5.4, we can therefore conclude that the trivial solution is unstable. Theorem 5.1 (see the remark) does not work in this case, since all the elements of the first approximation matrix vanish.

The proof of Theorem 5.4 will be carried out by *reductio ad absurdum*. Assume that the trivial solution is stable. Then $\forall \varepsilon > 0 \; \exists \delta(\varepsilon)$ such that $\|x(t)\| < \varepsilon$ if $t \geq 0$, whenever $\|x(0)\| < \delta(\varepsilon)$. Since ω_δ contains a point x_0 such that $V(x_0) > 0$ (see assumption a) of Theorem 5.4), let us take it for the initial point and denote $V(x_0) = \alpha$. For a sufficiently small t, the sign of the difference $V(x(t)) - V(x(0))$ is determined by the value $\left. \dfrac{dV}{dt} \right|_{t=0}$. But, according to b), we have $\left. \dfrac{dV}{dt} \right|_{t=0}$ $= W(x(0), 0) \geq \beta > 0$. Therefore $V(x(t)) > \alpha$ for sufficiently small t. Let us check that $V(x(t)) > \alpha$ for all $t > 0$. Assume that for some $t = t_1 > 0$ the function $V(x(t))$ again assumes the value α, i.e. $V(x(t_1)) = \alpha$. But then, on one hand, $V(x(t_1)) - V(x(0)) = 0$ while on the other,

$$V(x(t_1)) - V(x(0)) = \left. \frac{dV}{dt} \right|_{t=t^*} t_1 = W(x(t^*), t^*) t_1 \geq \beta t_1 > 0.$$

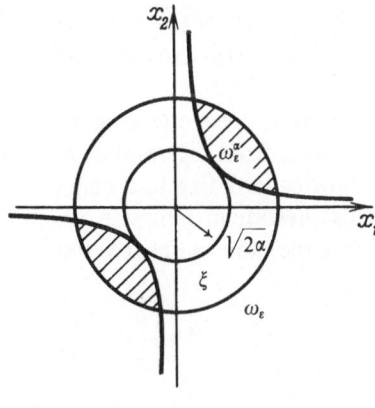

Fig. 12

This contradiction leads us to conclude that $V(x(t)) > \alpha$ for $t > 0$. Moreover, according to the stability assumption, $x(t) \in \omega_\varepsilon$ for $t \geq 0$. But then

$$V(x(t)) - V(x(0)) = W(x(t^{**}), t^{**}) t \geq \beta t.$$

This implies $V(x(t)) \to \infty$ as $t \to \infty$, while, on the other hand, $V(x(t))$ must be bounded in view of the condition $x(t) \in \omega_\varepsilon$. This contradiction concludes the proof of the theorem.

Remarks. 1. The main drawback of the approach developed in this section is that there is no sufficiently general constructive method for finding the function $V(x)$. Nevertheless, for a series of very important classes of differential systems, such a construction is possible.

2. One of these classes is the class of linear systems with constant coefficients. Theorem 5.1 on stability with respect to first approximation may be proved using Lyapunov functions for linear systems with constant coefficients, as is done, for example, in L. S. Pontrjagin's textbook [17].

3. For differential equations describing certain mechanical systems, the role of the Lyapunov function is played by the potential energy $V(x)$. The system itself is of the form $m\ddot{x} = -\operatorname{grad} V$. The equilibrium position for the system is determined by the condition $\operatorname{grad} V = 0$ (from the mathematical point of view the equilibrium position is a solution of the type $x = \bar{x} = \text{const}$; by an appropriate coordinate transformation, we can always transform the point $\bar{x}$ into the origin) i.e. the equilibrium position is a stationary point of potential energy. If the stationary point is a minimal point of potential energy, then the equilibrium position is stable. Indeed, in this case $V(x)$ is a positive definite function in a neighbourhood of the origin while the corresponding function $W(x, t)$, which equals $(\operatorname{grad} V, f) = -(\operatorname{grad} V)^2$, obviously also satisfies the assumptions of Theorem 5.2.

§4. The Study of Trajectories in a Neighbourhood of a Stationary Point

In a number of questions, it turns out to be necessary not only to study the point $(0, \ldots, 0)$ of phase space from the point of view of its stability, but also to find out what the trajectories look like in a neighbourhood of this point.

We shall not treat this question in complete generality, but limit ourselves to the case $n = 2$ (the phase plane) and to linear systems with constant coefficients (5.11):

$$\frac{dx}{dt} = Ax, \qquad A = \begin{bmatrix} a_{11} & a_{12} \\ a_{21} & a_{22} \end{bmatrix}. \tag{5.39}$$

This system possesses the trivial solution $x_1 = 0$, $x_2 = 0$. From the point of view of kinematics, this is a state, hence the point $(0, 0)$ of the phase plane corresponding to this solution is referred to as a *stationary point*.

Note that the phase trajectory of the system (5.39) may be viewed as an integral curve of the equation

$$\frac{dx_1}{dx_2} = \frac{a_{11}x_1 + a_{12}x_2}{a_{21}x_1 + a_{22}x_2}. \tag{5.40}$$

At the point $(0, 0)$, the right-hand side of equation (5.40) is discontinuous, i.e. the assumptions of the existence and uniqueness theorem no longer hold. The point $(0, 0)$ is, according to the terminology of Chapter 2, a singular point. Therefore, *a priori*, it is possible that no integral curves of equation (5.40) pass through the point $(0, 0)$, or that more than one integral curve or even an infinite number of integral curves pass through this point.

As will be shown, the form of the trajectory in a neighbourhood of the point $(0, 0)$ is determined, just as its stability properties, by the characteristic roots of the matrix A.

Consider the following cases.

a) Suppose the characteristic roots λ_1 and λ_2 are real, distinct and have the same sign, say $\lambda_2 < \lambda_1 < 0$. In this case, the solution of system (5.39) is of the form (see §7 in Chapter 3)

$$x = C_1 \alpha_{(1)} e^{\lambda_1 t} + C_2 \alpha_{(2)} e^{\lambda_2 t}, \tag{5.41}$$

where $\alpha_{(i)} = \begin{pmatrix} \alpha_{(i)1} \\ \alpha_{(i)2} \end{pmatrix}$ are certain constant columns, the eigenvectors of the matrix A corresponding to the eigenvalues λ_i $(i = 1, 2)$. The point $(0, 0)$ is, according to Theorem 5.1, asymptotically stable, and $x \to 0$ as $t \to \infty$. Let us investigate the character of the approximation in more detail. We have

$$\frac{dx_2}{dx_1} = \frac{C_1 \lambda_1 \alpha_{(1)2} e^{\lambda_1 t} + C_2 \lambda_2 \alpha_{(2)2} e^{\lambda_2 t}}{C_1 \lambda_1 \alpha_{(1)1} e^{\lambda_1 t} + C_2 \lambda_2 \alpha_{(2)1} e^{\lambda_2 t}}.$$

Hence we see that $C_1 \neq 0$ implies

$$\lim_{t \to \infty} \frac{dx_2}{dx_1} = \frac{\alpha_{(1)2}}{\alpha_{(1)1}},$$

i.e. all the integral curves, except the one corresponding to $C_1 = 0$, enter the point $(0,0)$ with a common tangent direction, whose equation is $x_2 = (\alpha_{(1)2}/\alpha_{(1)1})x_1$ (denote it by I). Note that the straight line I is itself one of the trajectories, namely the one which corresponds to $C_2 = 0$. The trajectory corresponding to $C_1 = 0$ is also a straight line and has the equation $x_2 = (\alpha_{(2)2}/\alpha_{(2)1})x_1$ (denoted by II). The lines I and II do not coincide, since the linear independence of the vectors $\alpha_{(i)}$ implies $\alpha_{(1)1}\alpha_{(2)2} - \alpha_{(2)1}\alpha_{(1)2} \neq 0$. The disposition of the straight lines I and II, and of the other trajectories, is shown schematically on Fig. 13. The arrows indicate the direction in which t increases.

If $\lambda_2 > \lambda_1 > 0$, then the disposition of the trajectories will remain exactly the same (if we let t tend to $-\infty$, we can carry out the same arguments as the ones above for $t \to \infty$). If the arrows still denote the direction in which t increases, then the direction of the arrows is now changed to the opposite ones by comparison to Fig. 13. The point $(0,0)$ is unstable in this case[3].

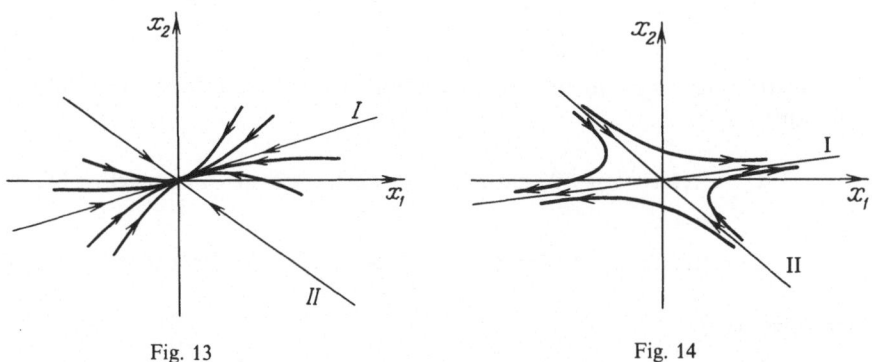

Fig. 13   Fig. 14

A stationary point corresponding to the case of real characteristic roots λ_1, λ_2, not equal to each other but having the same sign, is known as a *nodal point*. A nodal point is asymptotically stable for $\lambda_1 < 0$, $\lambda_2 < 0$ and unstable for $\lambda_1 > 0$, $\lambda_2 > 0$.

b) Suppose the characteristic roots λ_1 and λ_2 are real, different from each other and have different signs: $\lambda_1 > 0$, $\lambda_2 < 0$. The stationary point is unstable in this case. The representation (5.41) remains valid. It can be seen from (5.41) that two trajectories pass through the point $(0,0)$; one of them (denoted by I)

---

[3] We sometimes say that such a point is stable for $t \to -\infty$. The definition of stability for $t \to -\infty$ is quite similar to the definition of stability for $t \to \infty$ given in §1 of the present chapter, and theorems similar to the theorems in §§2 and 3 also hold.

corresponds to $C_2 = 0$ and has the equation $x_2 = (\alpha_{(1)2}/\alpha_{(1)1})x_1$, while the other (denoted by II) corresponds to $C_1 = 0$ and has the equation $x_2 = (\alpha_{(2)2}/\alpha_{(2)1})x_1$. But here the similarity with a nodal point ends. Along the trajectory II, $x \to 0$ as $t \to \infty$, while along trajectory I, $x \to 0$ as $t \to -\infty$ and the arrows indicating the direction in which t increases are directed away from the point $(0, 0)$. It is easy to see that the line I is an asymptote for $t \to \infty$ for all the trajectories, except II, since $\lim\limits_{t \to \infty} \dfrac{x_2}{x_1} = \dfrac{\alpha_{(1)2}}{\alpha_{(1)1}}$. The line II plays a similar role for $t \to -\infty$. The disposition of the trajectories is schematically shown on Fig. 14.

A stationary point corresponding to the case of real characteristic roots of different signs is known as a *saddle point*. The saddle point is an unstable stationary point.

The trajectories I and II passing through the saddle point are called *separatrices*.

c) Suppose the characteristic roots of the matrix A are complex. Since A is a real matrix, they will be conjugate complex numbers, i.e. $\lambda_1 = \lambda_2^* = \lambda$. The corresponding components of the eigenvectors $\alpha_{(i)}$ will also be conjugate, and since we are considering real solutions, the arbitrary constants must also be conjugate. Thus

$$x_1 = C\alpha_1 e^{\lambda t} + C^*\alpha_1^* e^{\lambda^* t}, \qquad x_2 = C\alpha_2 e^{\lambda t} + C^*\alpha_2^* e^{\lambda^* t}. \tag{5.42}$$

Substituting $\lambda = p + iq$ into this equation, we obtain (after a little manipulation)

$$x_1 = e^{pt}(2\alpha \cos qt - 2\beta \sin qt), \qquad x_2 = e^{pt}(2\gamma \cos qt - 2\delta \sin qt), \tag{5.43}$$

where

$$\alpha = \operatorname{Re}(C\alpha_1), \quad \beta = \operatorname{Im}(C\alpha_1), \quad \gamma = \operatorname{Re}(C\alpha_2), \quad \delta = \operatorname{Im}(C\alpha_2).$$

Hence we have

$$\varrho^2 = x_1^2 + x_2^2 = e^{2pt}[(2\alpha \cos qt - 2\beta \sin qt)^2 + (2\gamma \cos qt - 2\delta \sin qt)^2].$$

If $p = 0$ (the characteristic roots are purely imaginary), then x_1, x_2 and ϱ are periodic functions of t, of period $2\pi/q$. This means that a certain closed curve corresponds to every C on the phase plane. These curves do not intersect each other, since the uniqueness theorem holds for (5.40) everywhere except at the point $(0, 0)$ (see Fig. 15).

A more detailed study shows that the closed curves with which we are concerned are ellipses. Indeed, the determinant

$$\begin{vmatrix} 2\alpha & -2\beta \\ 2\gamma & -2\delta \end{vmatrix} \neq 0$$

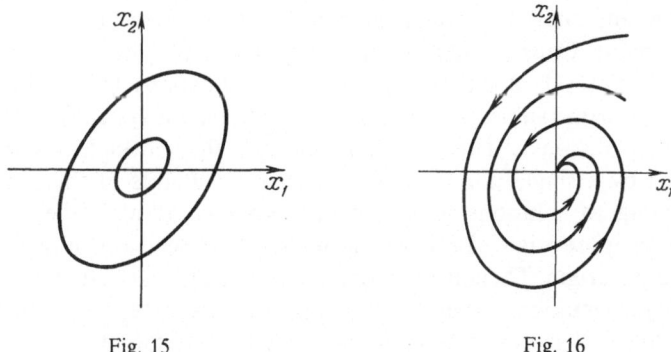

Fig. 15   Fig. 16

is nonzero, for otherwise a solution of the form $x_1 = ax_2$ (a is real) would exist. Since x_1 and x_2 are expressed by formula (5.42), it follows from the relation $x_1 = ax_2$, in view of the linear independence of $e^{\lambda t}$ and $e^{\lambda^* t}$, that $\alpha_1 = a\alpha_2$. The eigenvector $\begin{pmatrix} \alpha_1 \\ \alpha_2 \end{pmatrix}$ of the matrix A corresponding to the eigenvalue λ can be found from equation $(a_{11} - \lambda)\alpha_1 + a_{12}\alpha_2 = 0$, which, for $\alpha_1 = a\alpha_2$, yields $(a_{11} - \lambda)a = -a_{12}$; but this is possible only for real λ.

Thus (5.43) may be solved with respect to $\cos qt$ and $\sin qt$; setting the sum of their squares equal to one, we obtain

$$(b_{11}x_1 + b_{12}x_2)^2 + (b_{21}x_1 + b_{22}x_2)^2 = 1.$$

In the case $p \neq 0$, when t changes by values equal to the period, the function $\varrho(t)$ does not return to the previous value – this value increases or decreases according to the sign of p: $p > 0$ or $p < 0$.

On the phase plane we no longer obtain closed curves, but spiral-like ones. If $p < 0$, then $\varrho \to 0$ as $t \to \infty$ and the spiral winds into the point $(0,0)$, which is asymptotically stable (Fig. 16). In the case $p > 0$, we have a similar picture for $t \to -\infty$, while if t increases, then the spiral unwinds from the point $(0,0)$.

A stationary point corresponding to complex conjugate characteristic roots with a nonzero real part is said to be a focal point. When $p < 0$ the focus is asymptotically stable, while for $p > 0$ it is unstable.

A stationary point corresponding to purely imaginary characteristic roots is known as a vortex point. Centres are stable but not asymptotically stable stationary points (see §1). We shall not dwell on the case of multiple characteristic roots, see for example [22], nor on the case of vanishing characteristic roots.

Remarks. 1. The point $(0,0)$ was referred to as a stationary point for the linear system (5.39). Let us give a general definition of stationary points, from which it will be clear that a stationary point is not necessarily the origin of coordinates and that, for nonlinear systems, there may be more than one stationary point. Consider the nonlinear autonomous system (5.19) for $n=2$. Suppose $x_i = \bar{x}_i$ $(i=1,2)$ satisfies the system of equations $f_i(x_1, x_2) = 0$. Then, obviously, the same values $x_i = \bar{x}_i$ satisfy the differential system (5.19), since $\bar{x}_i$ does not depend on t. This solution describes the motionless state and is represented by a single point on the phase plane. Points of the phase plane corresponding to solutions of the form $x_i = \bar{x}_i = \text{const}$ are called *stationary* points. By an appropriate change of variables, we can send any of the stationary points into the origin. Then, if f_i is represented in the form (5.20), it follows that the first approximation system (5.22) coincides with (5.39).

According to Theorem 5.1, in the case $\mathrm{Re}\,\lambda \neq 0$, the stability or non-stability of the point $(0,0)$ of system (5.19) is guaranteed by the same assumptions on λ which guaranteed the stability or non-stability of the point $(0,0)$ for the first approximation system (5.22), i.e. the terms $R_{(2)i}$ do not influence the stability or non-stability of the point $(0,0)$. As to the disposition of the trajectories, a precise investigation see [17], § 30 shows that in case of a nodal point, a saddle point or a focal point of system (5.22) (in all these cases $\mathrm{Re}\,\lambda \neq 0$), the qualitative character of the disposition of the trajectories of the system (5.19) in a sufficiently small neighbourhood of the point $(0,0)$ will be the same. If at the point $(0,0)$ the system (5.22) has a vortex point, then, without a supplementary investigation of the terms $R_{(2)i}$, we cannot say anything about the disposition of the trajectories of system (5.19).

2. Our study of the disposition of trajectories in a neighbourhood of a stationary point gives certain information on the disposition of phase trajectories in the entire plane, but of course does not yield a complete solution of this complicated global problem.

In order to study the picture on the phase plane, or, as it is sometimes called, the phase portrait of system (5.19), it is important to have information not only about stationary points. One often meets closed trajectories (a closed trajectory corresponds to a periodic motion) which possess the property that in their neighbourhood there are no other closed trajectories and all the trajectories wind around this unique closed trajectory, which in this case is known as a limit cycle; or, conversely, unwind from it. Thus limit cycles can be either stable or unstable (on Fig. 17 we have pictured a stable limit cycle). The investigation of limit cycles is also important from the point of view of the existence of stable periodic regimes in physical systems.

Example. The system

$$\frac{dx_1}{dt} = -x_1 \left(\sqrt{x_1^2 + x_2^2} - a \right) - x_2, \qquad \frac{dx_2}{dt} = -x_2 \left(\sqrt{x_1^2 + x_2^2} - a \right) + x_1 \quad (5.44)$$

has the stationary point $x_1 = 0$, $x_2 = 0$ and the limit cycle $x_1^2 + x_2^2 = a^2$. It is easy to check the existence of this cycle by passing to polar coordinates

Fig. 17

$x_1 = \varrho \cos \theta$, $x_2 = \varrho \sin \theta$, in which the system (5.44) acquires the form

$$\frac{d\varrho}{dt} = -\varrho(\varrho - a), \qquad \frac{d\theta}{dt} = 1.$$

3. When the dimension n increases, new phenomena arise, for the study of which numerous methods have been developed.

The study of the phase portrait of the system of differential equations is one of the problems of the so-called qualitative theory of differential equations; see, for example, [15].

Chapter VI
Numerical Methods for the Solution of Ordinary Differential Equations

In practice, one does not usually succeed in finding the solution of ordinary differential equations in terms of elementary or special functions (notable exceptions are linear equations, considered in Chapter 3). At the same time, the extensive use of differential equations as mathematical models of a wide range of scientific problems requires the development of methods for their study which could be used to find numerical characteristics of the problem with guaranteed precision. Numerical solution methods turn out to be the most efficient here. Thanks to the rapid development of computers, numerical methods are now widely used in many branches of mathematics and its applications, in particular in the solution of ordinary differential equations. In recent years a large number of textbooks dealing with this question have appeared (see [18], [7], [13], [1]). This chapter will merely indicate the simplest problems relating to numerical methods for the solution of ordinary differential equations.

§1. Numerical Methods for Solving Initial Value Problems

1. Euler's method. In Chapter 2, when we studied the existence and uniqueness of the initial value problem's solution for the ordinary differential equation

$$\frac{dy}{dx} = f(x, y), \quad y(x_0) = y_0, \tag{6.1}$$

we noted that the method used to prove the existence theorem, namely the Euler algorithm, also gives an effective method for the numerical construction of an approximate solution of the problem. The Euler algorithm is the simplest numerical method for the solution of the initial value problem. Recall its main idea. The segment $[x_0, X]$ on which the solution of problem (6.1) is considered is divided into n parts by partition points

$$x_0, x_1, x_n = X; \quad x_i - x_{i-1} = h_i > 0, \ h_{(n)} = \max_i \{h_i\}.$$

and the Euler polygonal line is constructed; it is a piecewise linear function $\bar{y}_{(n)}(x)$ which is the solution of the initial value problem for equation (6.1) with a piecewise constant right-hand side:

$$\frac{d\bar{y}_{(n)}}{dx}=f(x_{i-1},\bar{y}_{(n)}(x_{i-1})), \quad x_{i-1}\leq x\leq x_i, \quad \bar{y}_{(n)}(x_0)=y_0. \tag{6.2}$$

In Chapter 2 we showed that the solution of Problem (6.2), being an approximate solution of the given problem (6.1), will not leave the domain D in which the function $f(x, y)$ satisfies the assumptions of the existence theorem. Further, studying other algorithms constructing approximate solutions of problem (6.1), we shall assume that the approximate solutions constructed according to them also remain within D, without involving ourselves in an estimation of the length $X-x_0$ of the segment where this takes place. Obviously, for $x_{i-1}\leq x\leq x_i$, the solution of problem (6.2) may be written in the form

$$\bar{y}_{(n)}(x)=\bar{y}_{(n)}(x_{i-1})+(x-x_{i-1})f(x_{i-1}, \bar{y}_{(n)}(x_{i-1})). \tag{6.3}$$

Formula (6.3) is an explicit scheme for the numerical construction of Euler's polygonal line, since it enables us to compute successively the values of the required function $\bar{y}_{(n)}(x)$ at all partition points x_i of the segment $[x_0, X]$. The geometric meaning of these constructions is quite simple. At each step (x_{i-1}, x_i) the motion along the corresponding integral curve of equation (6.1) is replaced by motion along the tangent to the integral curve at the point $x_{i-1},\bar{y}_{(n)}(x_{i-1})$. Note that for a sufficiently large step $h_{(n)i}$ and a large integration segment $[x_0, X]$ the final value $\bar{y}_{(n)}(X)$ may differ greatly from the value $y(X)$ of the original integral curve (see Fig. 18); however, it follows from Theorem 2.1 that for $h_{(n)}\to0$ the sequence of Euler polygonal lines $\{\bar{y}_{(n)}(x)\}$ converges to the original integral curve $y(x)$ of the initial value problem (6.1).

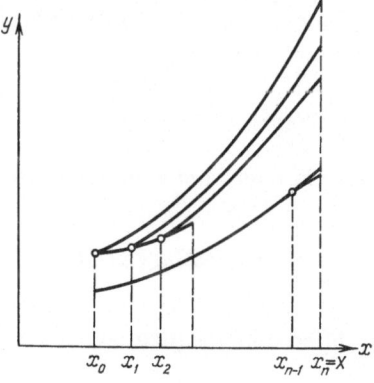

Fig. 18

Let us estimate how quickly the Euler method converges. Assume that the function $f(x, y)$ has continuous partial derivatives with respect to both arguments in a rectangle D where the solution of problem (6.1) exists and is unique. Then we have the estimates

$$|f(x, y)| < M, \quad \left|\frac{\partial f}{\partial x}(x, y)\right|, \quad \left|\frac{\partial f}{\partial y}(x, y)\right| < K. \tag{6.4}$$

Consider the difference $z(x)$ between the exact solution $y(x)$ of the given problem (6.1) and Euler's polygonal line $\bar{y}_{(n)}(x)$ satisfying the equation and the initial condition (6.2):

$$z(x) = y(x) - \bar{y}_{(n)}(x). \tag{6.5}$$

By (6.1), (6.2), (6.5) we obtain

$$\frac{dz}{dx} = f(x, y(x)) - f(x_{i-1}, \bar{y}_{(n)}(x_{i-1})), \tag{6.6}$$

$$z(x_0) = 0. \tag{6.7}$$

Rewrite equation (6.6) in the form

$$\frac{dz}{dx} = f(x, y(x)) - f(x, \bar{y}_{(n)}(x)) - \psi_n(x) \tag{6.8}$$

where (compare with (2.61))

$$\psi_n(x) = f(x_{i-1}, \bar{y}_{(n)}(x_{i-1})) - f(x, \bar{y}_{(n)}(x)). \tag{6.9}$$

In view of the conditions imposed on $f(x, y)$, we have

$$|\psi_n(x)| \le Kh_{(n)} + K|\bar{y}_{(n)}(x_{i-1}) - \bar{y}_{(n)}(x)| \le Kh_{(n)} + KMh_{(n)} \tag{6.10}$$

i.e.

$$|\psi_n(x)| \le Kh_{(n)}(1 + M). \tag{6.11}$$

Using (6.11) and (6.4) we can pass from equation (6.8) to the inequality

$$\left|\frac{dz}{dx}\right| \le K|z| + Kh_{(n)}(1 + M). \tag{6.12}$$

Hence by lemma 2.1 on differential inequalities,

$$|z(x)| \le h_n(1 + M)\, (e^{K(x - x_0)} - 1). \tag{6.13}$$

It can be seen from this formula that the expression $|z(x)|$ depends linearly on the length $h_{(n)}$ of the maximal step in the partition. This means that the approximate solution $\bar{y}_{(n)}(x)$ obtained by the Euler polygonal line method (6.3) converges to an exact solution of the original problem with precision of the first order. The exponential term in the estimate (6.13) characterizes the divergence of the integral curves (see Fig. 18).

Remark. The estimate of the error (6.13) is an estimate from above. For functions $f(x, y)$ with derivatives whose sign changes, it can turn out to be much larger than more precise asymptotic estimates, however, the order of precision with respect to $h_{(n)}$ will remain equal to one, but the coefficients of $h_{(n)}$ will be specified in a different way.

2. General notions of the theory of difference schemes. In the previous subsection we obtained, without much difficulty, an estimate of the speed of convergence of the simplest finite difference method, i.e. Euler's method. In order to study more complicated methods, possessing a higher order of convergence, we shall need some general notions from the theory of difference schemes.

For most numerical methods it is sufficient to indicate an algorithm for computing the approximate solution at a finite number of fixed points x_i of the segment $[x_0, X]$ on which we are solving problem (6.1). The set of points $\omega_n = \{x_i\}$ $(i = 0, 1, \ldots, n)$ where the approximate solution is investigated is called a net. The points of this set are the nodes of the net. The distance $h_i = x_i - x_{i-1}$ $(i = 1, \ldots, n)$ between successive nodes is the step of the net. A net may be non-uniform $h_i \neq \text{const}$ (in subsection 1 we considered the Euler method for non-uniform nets) and uniform $h_i = h = \text{const}$. Obviously, in the latter case, the step of the net on the segment $[x_0, X]$ equals $h = [X - x_0]/n$. Further, unless the converse is explicitly stated, we shall consider uniform nets.

A function of a discrete argument $y(x_i)$, defined only at nodes of the net, is called a net function. The values of the net function $y_{(h)}$ on ω_h will be denoted by $y_{(h),i}$, $i = 0, 1, \ldots, n$.

Finite difference methods for the solution of problems of differential equations consist in replacing the given differential problem for functions of a continuous argument by an algebraic problem for the net function; namely, by a system of algebraic equations expressing the relationship between the values of the net functions, the given supplementary conditions and the right-hand sides of the equations at the nodes x_i of the net ω_h.

Suppose the differential problem is of the form

$$Ly = \varphi(x). \tag{6.14}$$

Here the symbol L will denote not only a specification of the given equation, but also that of the supplementary conditions, say initial conditions, written in a definite order. $\varphi(x)$ denotes the right-hand side of the equation and the right-hand side of the supplementary conditions, written in the corresponding order. Thus problem (6.1) may be written in the form

$$Ly = \left\{ \begin{array}{c} \dfrac{d}{dx}\, y - f(x,y) \\ y(x_0) \end{array} \right\} = \left\{ \begin{array}{c} 0 \\ y_0 \end{array} \right\}. \tag{6.15}$$

The corresponding difference problem can be written as

$$L_h y_{(h)} = \varphi_{(h)}, \tag{6.16}$$

where L_h denotes the specification of the difference equation and the corresponding supplementary conditions, while $\varphi_{(h)}$ is the input data of the problem, i.e. the values of the net function $f_{(h)}$ in the right-hand side of the difference equations, together with the right-hand sides of the supplementary conditions. Equation (6.22) given below is an example of this type of notation for Euler's scheme.

The problem of determining the net functions $y_{(h)}$ must be set so that when the step h of the net tends to zero, the net's functions converge, in a certain sense, to the exact solution of the original problem (6.14).

In order to determine the convergence of a family of net functions to the solution of the given problem in the space $\{v_{(h)}\}$ of net functions, it is necessary to define the distance between different functions as the norm of their difference. The notion of norm in the space of net functions may be introduced in different ways. Most often, one uses the so-called uniform Chebyshev norm, defined by the expression

$$\|v_{(h)}\| = \max_i |v_{(h)i}| \quad (i = 0, 1, \ldots, n). \tag{6.17}$$

In many cases one uses the mean quadratic (Hilbert) norm

$$\|v_{(h)}\|_{l_2} = \left(\sum_{i=1}^{n} v_{(h)i}^2 \varrho_i h_i \right)^{1/2} \tag{6.18}$$

where ϱ_i are given weight coefficients. In certain cases other norms are used, e. g. energy norms.

We shall say that the family of net functions $\{y_{(h)}\}$ converges to the exact solution $y(x)$ of the given problem if

$$\lim_{h \to 0} \|y_{(h)} - [y]_h\| = 0, \tag{6.19}$$

where $[y]_h$ denotes the value of the function $y(x)$ on the net ω_h. If, moreover,

$$\| y_{(h)} - [y]_h \| < Ch^k \qquad (6.20)$$

where C and k are certain constants, which do not depend on h, then we shall say that the convergence is of order k in the corresponding norm. In subsection 1 it was shown that the family of net functions $\{y_{(h)}\}$ obtained by the Euler method converged to the exact solution in the uniform norm with order 1. However, in the general case, such an investigation is quite difficult. Further we shall show that the convergence property of a difference scheme may be reduced to two other properties, approximation and stability, which are easier to check.

Now let us pass to the definition of a difference scheme's order of approximation. To do this, we must introduce the norm of $\varphi_{(h)}$. As was pointed out above, $\varphi_{(h)}$ is the right-hand side of the equation, i.e. the net function $f_{(h)}$ together with the supplementary conditions for the net function $y_{(h)}$. They may be initial values or boundary values. Let us number them is some definite order. Suppose the corresponding right-hand sides will be $g_1, \ldots, g_q$.

Let us denote $\| y_{(h)} \|_0 = \max |g_i|$ and introduce the norm $\| \varphi_{(h)} \|_1$ by putting

$$\| \varphi_{(h)} \|_1 = \max \{ \| f_{(h)} \|, \quad \| y_{(h)} \|_0 \}.$$

If we replace the original differential problem (6.14) by the difference problem (6.16), the exact solution of the given problem in general will not satisfy the difference problem. We shall say that the difference scheme approximates the given problem with order k if

$$\| L_h[y]_h - \varphi_{(h)} \|_1 < Ch^k, \qquad (6.21)$$

where C and k are certain constants which do not depend on h.

Euler's scheme (6.3) means that the given differential problem (6.1) was replaced by the difference problem

$$L_h y_{(h)} - \left\{ \begin{array}{c} \dfrac{y_{(h)i} - y_{(h)i-1}}{h} - f_{(h)i-1} \\[2mm] y_{(h)0} \end{array} \right\} - \left\{ \begin{array}{c} 0 \\ y_0 \end{array} \right\}, \qquad (6.22)$$

where $f_{(h)i} = f(x_i, y_{(h),i})$.

Let us determine the order of approximation of this scheme, assuming that the right-hand side $f(x, y)$ of equation (6.1) satisfies the same conditions as in subsection 1 (see page 156). Then, obviously, the exact solution $y(x)$ of the given problem will be a function which is continuously differentiable twice. This allows us to write out the expansion

$$y(x_i + h) = y(x_i) + hy'(x_i) + \frac{h^2}{2} y''(x_i + \theta h) \qquad (0 \le \theta \le 1). \qquad (6.23)$$

By (6.23), if we substitute the exact solution $y(x)$ of problem (6.1) into the left-hand side of equation (6.22), we get

$$y'(x_{i-1}) + \frac{h}{2} y''(x_{i-1} + \theta h) - f(x_{i-1}, y(x_{i-1})) = \frac{h}{2} y''(x_{i-1} + \theta h), \quad (6.24)$$

First and third summands in the left-hand side of (6.24) cancel out each other, since $y(x)$ is the exact solution of equation (6.1). The solution of the difference problem (6.22) satisfies the same initial condition as the solution of (6.1). This implies that Euler's scheme gives a first order approximation for problem (6.1).

Remark. The first order of approximation in Euler's scheme is related to the fact that the first derivative $y'(x_i)$ in this method is approximated by the difference quotient $(y_i - y_{i-1})/h$. It is easy to see that, choosing other approximate formulas for the first derivative, we can raise the order of approximation. Indeed, let us replace the first derivative $y'(x_i)$ by symmetric difference quotient at successive points

$$\frac{y_{i+1} - y_{i-1}}{2h}. \quad (6.25)$$

Assuming that the third derivatives of the solution are continuous (which will be the case if the appropriate smoothness conditions are imposed on the function $f(x, y)$ in the right-hand side of equation (6.2)), we shall have

$$y(x_i + h) = y(x_i) + hy'(x_i) + \frac{h^2}{2} y''(x_i) + \frac{h^3}{6} y'''(x_i + \theta_1 h),$$

$$y(x_i - h) = y(x_i) - hy'(x_i) + \frac{h^2}{2} y''(x_i) - \frac{h^3}{6} y'''(x_i + \theta_2 h),$$

$$(6.26)$$

Then, approximating problem (6.1) by the difference scheme

$$L_h y_{(h)} = \left\{ \begin{array}{c} \dfrac{y_{(h)i+1} - y_{(h)i-1}}{2h} - f_{(h)i} \\[2mm] y_{(h)0} \\[2mm] y_{(h)1} \end{array} \right\} = \left\{ \begin{array}{c} 0 \\ y_0 \\ y_1 \end{array} \right\} \quad (6.27)$$

and carrying out arguments similar to the previous ones, we see that the difference equation (6.27) approximates the given equation (6.1) with second order precision. Note that the scheme (6.27) requires knowledge of the net function not only at the zeroth, but at the first node. In order to determine the order of approximation of this condition, let us again appeal to a Taylor expansion of the exact solution

$$y(x_0+h)-y_1=y(x_0)+hy'(x_0)+\frac{h^2}{2}\,y''(x_0+\theta h)-y_1$$

$$=y_0+hf(x_0,y_0)+\frac{h^2}{2}\,y''(x_0+\theta h)-y_1. \tag{6.28}$$

It follows from (6.28) that, if we choose

$$y_1=y_0+hf(x_0,y_0), \tag{6.29}$$

then the second initial condition (6.27) will also be approximated with second order. It therefore follows that if the supplementary condition (6.29) holds, then the difference scheme (6.27) is a second order approximation of the given problem.

Now let us turn to the definition of the next main notion of the theory of different schemes i.e. the notion of stability of difference schemes. We begin with a simple example.

Consider problem (6.1) for the first order linear homogeneous equation with constant coefficient

$$\frac{dy}{dx}=\alpha y, \quad y(x_0)=y_0, \tag{6.30}$$

where α is the given constant. For problem (6.30), let us construct a difference scheme which is an obvious generalization of the difference schemes (6.22) and (6.27)

$$\sigma(y_{(h)i+1}-y_{(h)i})+(1-\sigma)\,(y_{(h)i}-y_{(h)i-1})-h\alpha y_{(h)i}=0$$

$$y_{(h)0}=y_0, \quad y_{(h)1}=y_1, \tag{6.31}$$

where σ is a certain constant satisfying the condition $0\le\sigma\le1$, while the value y_1 is determined by condition (6.29). Obviously, formula (6.31) gives a one parameter family of difference schemes depending on the parameter σ. For $\sigma=1$, we obtain the scheme (6.22), for $\sigma=1/2$ we get the symmetric scheme (6.27). Suppose the initial conditions of the scheme (6.31) are given with a certain error ε. Then we obtain a new net function $\bar{y}_{(h)}$ which is the solution of the problem

$$\sigma(\bar{y}_{(h)i+1}-\bar{y}_{(h)i})+(1-\sigma)\,(\bar{y}_{(h)i}-\bar{y}_{(h)i-1})-h\alpha\bar{y}_{(h)i}=0$$

$$\bar{y}_{(h)0}=y_0+\varepsilon_0; \quad \bar{y}_{(h)1}=y_1+\varepsilon_1. \tag{6.32}$$

Let us see how the error in the initial conditions influences the solution of problem (6.31). Define the error of the solution by putting

$$\delta y_i=\bar{y}_{(h)i}-y_{(h)i}$$

Since the problems (6.31), (6.32) are linear, we have

$$\sigma(\delta y_{i+1} - \delta y_i) + (1 - \sigma)(\delta y_i - \delta y_{i-1}) - h\alpha\delta y_i = 0, \qquad (6.33)$$

hence, for the error, we obtain the problem

$$\sigma\delta y_{i+1} + (1 - 2\sigma - h\alpha)\delta y_i - (1 - \sigma)\delta y_{i-1} = 0$$
$$\delta y_0 = \varepsilon_0, \quad \delta y_1 = \varepsilon_1. \qquad (6.34)$$

Problem (6.34) is a particular case of a linear difference problem with constant coefficients

$$a_0 y_k + a_1 y_{k-1} + \ldots + a_q y_{k-q} = 0$$
$$y_0 = b_0, \quad y_1 = b_1, \ldots, y_{q-1} = b_{q-1}. \qquad (6.35)$$

Similary to linear differential equations with constant coefficients, a particular solution of the difference equation (6.35) may be searched for in the form $y_k = \lambda^k$, where λ is an unknown constant which is to be determined. Substituting the desired form of the solution into equation (6.35), we obtain the following characteristic equation, which is an algebraic equation of the q-th degree:

$$a_0 \lambda^q + a_1 \lambda^{q-1} + \ldots + a_q = 0. \qquad (6.36)$$

If all the roots $\lambda_p(p=1,\ldots,q)$ of equation (6.36) are simple, then we obtain q distinct particular solutions

$$y_k^{(p)} = \lambda_p^k \ (p=1,\ldots,q) \qquad (6.37)$$

of equation (6.35); its general solution, by linearity, will be expressed in the form

$$y_k = \sum_{p=1}^{q} C_p \lambda_p^k. \qquad (6.38)$$

It is easy to prove that the solutions of (6.37) are linearly independent in the case of simple root λ_p and constitute a fundamental system. We shall not dwell on this question see [9]. Determining the constants C_p in (6.38) from the initial conditions (6.35), we obtain the solution of problem (6.35).

Let us apply this general method to find the solution of problem (6.34). The corresponding characteristic equation is of the form

$$\sigma\lambda^2 + (1 - 2\sigma - h\alpha)\lambda - (1 - \sigma) = 0, \qquad (6.39)$$

while its roots are determined by the expressions

$$\lambda_{1,2} = -\frac{1-2\sigma-h\alpha}{2\sigma} \pm \sqrt{\frac{(1-2\sigma-h\alpha)^2}{4\sigma^2} + \frac{1-\sigma}{\sigma}}. \tag{6.40}$$

It is easy to see that, for sufficiently small h,

$$\lambda_1 = 1 + h\alpha + O(h^2), \quad \lambda_2 = 1 - \frac{1}{\sigma} + O(h). \tag{6.41}$$

Here the symbol $O(h^k)$ denotes an expression whose absolute value is less than Ch^k, where C is a constant which is independent of h.

It therefore follows that we have the estimate $|\lambda_2| > 1$ (if $\sigma < 1/2$) for the second root in (6.41). Therefore, the particular solution of the difference equation (6.34) corresponding to this root

$$\delta y_i^{(2)} = \lambda_2^i \tag{6.42}$$

has an absolute value which increases unboundedly (exponentially) as i tends to infinity. In this situation, the particular solution corresponding to the first root, for any value of σ ($0 \le \sigma \le 1$) remains bounded as $i \to \infty$. Indeed, since $i = (x_i - x_0)/h$, where $x_i \le X$, elementary manipulations yield

$$\delta y_i^{(1)} = \lambda_1^i = [1 + h\alpha + O(h^2)]^{(x_i - x_0)/h} = e^{\alpha(x_i - x_0)} + O(h), \tag{6.43}$$

which proves the uniform boundedness of $\delta y_i^{(1)}$ on the interval $x_0 \le x_i \le X$.

Let us look for the solution of problem (6.34) in the form

$$\delta y_i = C_1 \delta y_i^{(1)} + C_2 \delta y_i^{(2)}. \tag{6.44}$$

The constants C_1 and C_2 are determined from the initial conditions of problem (6.34). Simple computations yield

$$C_1 = (1 - \sigma)\varepsilon_0 + \sigma\varepsilon_1 + O(h), \tag{6.45}$$

$$C_2 = \sigma(\varepsilon_0 - \varepsilon_1) + O(h). \tag{6.46}$$

In order to obtain a numerical solution of the initial value problem (6.30) for any fixed $x \in [x_0, X]$ by using the difference scheme, carry out the number of steps i necessary to get $x_i = x_0 + ih \ge x$. Therefore, $|\lambda_2^i| = |\lambda_2^{(x_i - x_0)/h}| \ge |\lambda_2^{(x - x_0)/h}|$ grows exponentially with $h \to 0$. Thus it follows from (6.44) that the solution of problem (6.34) increases unboundedly because of the term $C_2 \delta y_i^{(2)}$. For $\varepsilon_0 = \varepsilon_1$, the expression C_2 tends to zero as $h \to 0$, but since C_2 decreases with order no greater than that of a polynomial, it follows that $C_2 \lambda_2^i$ (and hence the entire solution) will still be unbounded. Thus even a small error in the initial conditions can lead to an increasing error in the solution of problem (6.31) when the step in the difference scheme is decreased. The scheme (6.31) for $\sigma < 1/2$ is called non-stable with respect to the initial data, by analogy with the notion introduced in Chapter 5. In a similar way, it can be shown that the given scheme is non-stable with respect to the right-hand side for the nonhomogeneous equation (6.30) as well.

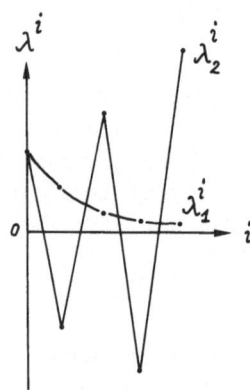

Fig. 19

Figure 19 shows the dependence of $\delta y_i^{(2)}$ and $\delta y_i^{(1)}$ on i for negative values of the parameter α. Since $\lambda_2 < 0$ and $|\lambda_2| > 1$, the absolute value of the expression $\delta y_i^{(2)}$ increases unboundedly when i tends to infinity and assumes negative and positive values alternately.

Now let us give a rigorous definition of a stable difference scheme. Consider the difference scheme (6.16) and the corresponding perturbed problem

$$L_h \bar{y}_{(h)} = \varphi_{(h)} + \delta \varphi_{(h)}. \tag{6.47}$$

Then $\delta \varphi_{(h)}$ is called a perturbation of the input data. The error of the solution is still, by definition, the expression $\delta y_{(h)} = \bar{y}_{(h)} - y_{(h)}$. The norm of the terms $\delta \varphi_{(h)}$ and $\varphi_{(h)}$ may be estimated as follows

$$\|\delta \varphi_{(h)}\|_1 = \max \{ \|\delta f_{(h)}\|, \|\delta y_{(h)}\|_0 \}. \tag{6.48}$$

In the sequel, we shall consider the case when the perturbed problem (6.47) has a unique solution for sufficiently small perturbations: i.e. there exist such δ_0 and h_0 that $\|\delta \varphi_{(h)}\|_1 < \delta_0$ and $h < h_0$ imply that problem (6.47) possesses a unique solution.

Definition. The scheme (6.16) is said to be *stable with respect to the initial data and to the right-hand side of the equation* (or simply stable) if there exists a constant h_0 such that $h < h_0$ implies that the norm of the error of the solution satisfies the inequality

$$\|\delta y_{(h)}\| < C\|\delta\varphi_{(h)}\|_1,\tag{6.49}$$

where the constant C does not depend on h.

The above definition obviously means that the solution of the difference scheme (6.16) depends continuously on the input data of the problem i.e. the initial data and the right-hand side of the equation: a small change in the input data brings about a small change in the solution.

It is easy to verify that the scheme (6.31) for $\sigma > 1/2$ is stable with respect to initial data. Stability with respect to initial data may be checked by setting equal to zero the $\delta f_{(h)}$-component in the expression for $\delta\varphi_{(h)}$ (as it was in our example). Indeed, as we proved previously, $\delta y_i^{(1)}$ is uniformly bounded. As for $\delta y_i^{(2)}$, we see from (6.41) that $\sigma > 1/2$ implies the inequality $|\lambda_2| < 1$ and therefore $\delta y_i^{(2)}$ is also uniformly bounded, and even infinitely small when h tends to zero. Hence (6.44), (6.45) and (6.46) imply

$$\|\delta y_{(h)}\| < C\varepsilon\tag{6.50}$$

where

$$\varepsilon = \max\{|\varepsilon_0|, |\varepsilon_1|\},\tag{6.51}$$

which proves our statement.

Now let us prove the stability of the Euler scheme (6.22). The corresponding perturbed problem will be of the form

$$L_h\bar{y}_{(h)} = \begin{cases} \dfrac{\bar{y}_{(h)i} - \bar{y}_{(h)i-1}}{h} - f(x_{i-1}, \bar{y}_{(h)i-1}) = \delta f_{(h)i-1}. \\[2mm] \bar{y}_{(h)0} = y_0 + \varepsilon_0. \end{cases}\tag{6.52}$$

We shall assume that the solution $\bar{y}_{(h)}$ of this problem belongs to the domain D in which inequalities (6.4) hold. For the solution error $\delta y_{(h)}$ of the difference problem (6.52) we obviously have

$$L_h\delta y_{(h)} = \begin{cases} \dfrac{\delta y_{(h)i} - \delta y_{(h)i-1}}{h} - f(x_{i-1}, \bar{y}_{(h)i-1}) + f(x_{i-1}, y_{(h)i-1}) = \delta f_{(h)i-1}, \\[2mm] \delta y_{(h)0} = \varepsilon_0. \end{cases}\tag{6.53}$$

Here

$$|\delta f_{(h)i-1}| \le \|\delta f_{(h)}\|, \quad |\varepsilon_0| = \|\delta y_{(h)}\|_0.\tag{6.54}$$

In view of the estimates (6.4), relation (6.53) implies

$$|\delta y_{(h)i}| \le |\delta y_{(h)i-1}| + hK|\delta y_{(h)i-1}| + h\|\delta f_{(h)}\| = (1 + hK)|\delta y_{(h)i-1}| + h\|\delta f_{(h)}\| \le \dots$$

$$\dots \le (1 + hK)^2|\delta y_{(h)i-2}| + (1 + hK)h\|\delta f_{(h)}\| + h\|\delta f_{(h)}\|.\tag{6.55}$$

Applying this process of successive estimates, we shall obtain after the i-th step

$$|\delta y_{(h)i}| \leq (1+hK)^i |\varepsilon_0| + h\{(1+hK)^{i-1} + \ldots + 1\} \|\delta f_{(h)}\|$$

$$\leq (1+hK)^i \|\delta y_{(h)}\|_0 + \frac{1}{K}(1+hK)^i \|\delta f_{(h)}\|. \tag{6.56}$$

In view of (6.48) and the definition of Chebyshev norm, this yields

$$\|\delta y_{(h)}\| \leq (1+hK)^n \left(\|\delta y_{(h)}\|_0 + \frac{1}{K} \|\delta f_{(h)}\| \right). \tag{6.57}$$

Since $X - x_0 = nh$, the inequality $(1+hK)^{(X-x_0)/h} \leq e^{K(X-x_0)}$ finally implies

$$\|\delta y_{(h)}\| \leq e^{K(X-x_0)} \left(\|\delta y_{(h)}\|_0 + \frac{1}{K} \|\delta f_{(h)}\| \right) < C \|\delta \varphi_{(h)}\|_1 \tag{6.58}$$

for any h; this proves the stability of the difference schemes (6.53).

The notions of order of approximation and stability of a difference scheme introduced above, play an essential role in the following theorem, which is the fundamental one in the study of the convergence of difference schemes.

Theorem 6.1. *If the difference scheme (6.16) is stable and approximates problem (6.14) with order k, then the solution $y_{(h)}$ converges as $h \to 0$ to the solution $y(x)$ of the differential problem and we have the estimate*

$$\|y_{(h)} - [y]_h\| < Ch^k, \tag{6.59}$$

where the constant C does not depend on h.

Proof. Denote the difference of values between the net functions $y_{(h)}$ and $[y]_h$ by $\delta y_{(h)}$:

$$\delta y_{(h)} = y_{(h)} - [y]_h. \tag{6.60}$$

In view of the definition of the order of approximation,

$$L_h[y]_h = \varphi_{(h)} + \delta \varphi_{(h)}, \tag{6.61}$$

where $\|\delta \varphi_{(h)}\|_1 < C_1 h^k$ and by (6.49), it follows from the stability of the difference scheme (6.16) that

$$\|\delta y_{(h)}\| < C \|\delta \varphi_{(h)}\|_1 < CC_1 h^k. \tag{6.62}$$

Retaining the notation C for the product of constants CC_1, we obtain relation (6.59). The theorem is proved.

It was shown previously that Euler's scheme is stable and approximates problem (6.1) with order 1. It follows from the theorem proved here that the family of solutions $y_{(h)}$ obtained by the Euler scheme converges (as $h \to 0$) to the exact solution of problem (6.1), also with order 1. This fact was established in subsection 1 by means of direct estimates.

3. The Runge-Kutta method. As we have already pointed out, Euler's scheme converges only with order 1; thus, to get sufficient precision in using this scheme, it is necessary to carry out computations with a very small step, which brings about a considerable increase of computation time. It is therefore natural to try to find schemes which possess higher orders of convergence. One of the classes of such schemes are the schemes of the Runge-Kutta method. This method is based on the following considerations. Consider the identity $y'(x) = f(x, y(x))$. According to the Lagrange formula on finite increments on the interval $[x_{i-1}, x_i]$ there exists a point x^* such that

$$y(x_i) - y(x_{i-1}) = (x_i - x_{i-1})y'(x^*) = hf(x^*, y(x^*)). \tag{6.63}$$

However, we do not know the value of x^*. In Euler's scheme, we put $x^* = x_{i-1}$ and this only yielded a precision of the first order. The main idea of the Runge-Kutta method is to introduce a number of supplementary parameters in the difference scheme, which make the approximate determination of the value of x^* more precise.

Once again we shall consider the initial value problem (6.1)

$$\frac{dy}{dx} = f(x, y), \qquad y(x_0) = y_0. \tag{6.1}$$

The smoothness conditions imposed on the function $f(x, y)$ will be stated later.

To problem (6.1) we assign the difference scheme

$$L_h y_{(h)} = \left\{ \begin{aligned} &\frac{y_{(h)i} - y_{(h)i-1}}{h} - (p_1 K_1 + \ldots + p_l K_l) \\ &y_{(h)0} \end{aligned} \right\} = \left\{ \begin{aligned} 0 \\ y_0 \end{aligned} \right\}, \tag{6.64}$$

where

$$K_1 = f(x_{i-1}, \ y_{(h)i-1}),$$
$$K_2 = f(x_{i-1} + \alpha_1 h, \ y_{(h)i-1} + \alpha_1 h K_1),$$
$$\cdots\cdots\cdots\cdots\cdots\cdots\cdots\cdots\cdots\cdots\cdots\cdots\cdots\cdots \tag{6.65}$$
$$K_l = f(x_{i-1} + \alpha_{l-1} h, \ y_{(h)i-1} + \alpha_{l-1} h K_{l-1}),$$

and $p_1, \ldots, p_l, \alpha_1, \ldots, \alpha_{l-1}$ are certain parameters whose choice will guarantee the required order of approximation of the scheme (6.64) for the solution of problem (6.1).

In the particular case when $l=1$ and $p=1$, scheme (6.64) becomes the Euler scheme (6.22) (which gives an approximation of order 1). Let us show that, for $l=2$, we can choose the parameters p_1, p_2 and α_1 so that scheme (6.64) will have an order of approximation equal to 2. We shall assume here that the function $f(x, y)$ has continuous partial derivatives up to the second order with respect to both variables in D which is sufficient for the existence of a continuous third derivative of the solution. Then, substituting the solution of problem (6.1) into scheme (6.64) for $l=2$ and developing the solution in a Taylor series, we obtain

$$
\begin{aligned}
y(x_i) - y(x_{i-1}) &= hy'(x_{i-1}) + \frac{h^2}{2} y''(x_{i-1}) + \frac{h^3}{6} y'''(x^*) \\
&= h[p_1 f(x_{i-1}, y(x_{i-1})) \\
&\quad + p_2 f(x_{i-1} + \alpha_1 h, \ y(x_{i-1}) + \alpha_1 hf(x_{i-1}, \ y(x_{i-1})))]. \quad (6.66)
\end{aligned}
$$

Using the expansions

$$
\begin{aligned}
&f(x_{i-1} + \alpha_1 h, \ y(x_{i-1}) + \alpha_1 hf(x_{i-1}, \ y(x_{i-1}))) \\
&= f(x_{i-1}, \ y(x_{i-1})) + \alpha_1 h \frac{\partial f}{\partial x}(\cdot) + \alpha_1 hf(\cdot) \frac{\partial f}{\partial y}(\cdot) + O(h^2) \quad (6.67)
\end{aligned}
$$

and the obvious relations

$$
y''(x_{i-1}) = \frac{\partial f}{\partial x}(\cdot) + \frac{\partial f}{\partial y}(\cdot) y'(x_{i-1}) = \frac{\partial f}{\partial x}(\cdot) + f(\cdot) \frac{\partial f}{\partial y}(\cdot), \quad (6.68)
$$

let us rewrite (6.66) in the form

$$
y'(x_{i-1}) - (p_1 + p_2) f(x_{i-1}, \ y(x_{i-1})) + hy''(x_{i-1}) \left(\tfrac{1}{2} - \alpha_1 p_2\right) = O(h^2). \quad (6.69)
$$

Therefore, whenever

$$
p_1 + p_2 = 1, \quad \alpha_1 p_2 = \tfrac{1}{2}, \quad (6.70)
$$

scheme (6.64) for $l=2$ is a second order approximation. Here α_1 may be chosen arbitrarily. Schemes (6.64) with $\alpha_1 = 1$ or $\alpha_1 = 1/2$ are most widely used. Note that it is impossible to increase the order of approximation of schemes (6.64) for $l=2$ by choosing α_1.

When $\alpha_1 = 1$, it follows from (6.70) that $p_1 = p_2 = 1/2$ and scheme (6.64) acquires the form

$$
\begin{aligned}
&\frac{y_{(h)i} - y_{(h)i-1}}{h} - \frac{1}{2} [f(x_{i-1}, \ y_{(h)i-1}) \\
&\quad + f(x_{i-1} + h, \ y_{(h)i-1} + hf(x_{i-1}, \ y_{(h)i-1}))] = 0. \quad (6.71)
\end{aligned}
$$

The geometric interpretation of this formula is clear from Fig. 20. First, by Euler's method, we find the point $\bar{y}_i = y_{(h)i-1} + hf(x_{i-1}, y_{(h)i-1})$, then find the mean value of the slope of the tangent to the integral curve for the step h: $\tan \alpha = \frac{1}{2}(y'_{(h)i-1} + \bar{y}'_i)$ and use it to specify the value of $y_{(h)i}$. Similar considerations are easily carried out in the case $\alpha = 1/2$. Such schemes are usually called "predictor-corrector" schemes.

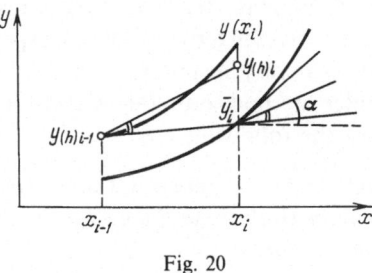

Fig. 20

We have considered schemes of approximation order two. Similar considerations are valid for higher order ($l > 2$) schemes. It is then necessary to impose higher smoothness requirements on the function $f(x, y)$. In practice, fourth order schemes are used most widely. In most standard programmes in computer software, the following scheme is used:

$$L_h y_{(h)} = \left\{ \begin{array}{c} \dfrac{y_{(h)i} - y_{(h)i-1}}{h} - \dfrac{1}{6}(K_1 + 2K_2 + 2K_3 + K_4) \\[2mm] y_{(h)0} \end{array} \right\} = \left\{ \begin{array}{c} 0 \\[2mm] y_0 \end{array} \right\}, \qquad (6.72)$$

where

$$K_1 = f(x_{i-1}, y_{(h)i-1}),$$

$$K_2 = f\left(x_{i-1} + \frac{h}{2}, y_{(h)i-1} + \frac{h}{2}K_1\right),$$

$$K_3 = f\left(x_{i-1} + \frac{h}{2}, y_{(h)i-1} + \frac{h}{2}K_2\right),$$

$$K_4 = f(x_{i-1} + h, y_{(h)i-1} + hK_3).$$

$$(6.73)$$

It is easy to check that this scheme gives an approximation of order four.

By theorem 6.1 if follows that, in order to determine the order of convergence of Runge-Kutta schemes, it is sufficient to prove their stability. This proof may be carried out in a manner completely similar to that of the Euler scheme's stability. Indeed, all schemes of types (6.64) are of the form

$$
L_h y_{(h)} = \left\{ \begin{array}{c} \dfrac{y_{(h)i} - y_{(h)i-1}}{h} - G(x_{i-1}, y_{(h)i-1}) \\[2mm] y_{(h)0} \end{array} \right\} = \left\{ \begin{array}{c} 0 \\ y_0 \end{array} \right\}, \tag{6.74}
$$

where the function $G(x, y)$ is a linear combination of the functions $f(x, y)$ at intermediate values of the argument. Hence it is easy to express estimates for the function $G(x, y)$ and its derivatives in terms of the corresponding estimates of the function $f(x, y)$ and its derivatives; the latter having been obtained in the proof of the stability of the Euler method, our statement is now established.

In particular, we have the following

Theorem 6.2. *If the function $f(x, y)$ possesses continuous partial derivatives of the fourth order in the domain D, then the Runge-Kutta scheme (6.72) converges with fourth degree of approximation, i.e.*

$$
\| y_{(h)} - [y]_h \| < Ch^4. \tag{6.75}
$$

Note in conclusion that the Runge-Kutta scheme can also be computed with a variable step. Beginning with any index i, we can increase or decrease the next step of the net.

Remark. In this section we have considered numerical methods for solving the Cauchy initial value problem (6.1) for one scalar equation of the first order. The methods considered here can be carried over without important changes to the case of initial value problems for normal systems of first order equations (see, for example, [18]).

§2. Boundary Value Problems

In this section we shall consider the simplest numerical methods for solving boundary value problems for ordinary differential equations. The formulation of such problems and the general properties of their solutions were discussed in Chapter 4.

1. The shooting method. The main idea of this method consists in reducing the solution of the given boundary value problem to an iterated solution of auxiliary Cauchy problems for the given differential equation. Let us illustrate this idea on the example of a boundary value problem on the segment $[0, l]$ for the second order equation

$$
\frac{d^2 y}{dx^2} = f\left(x, y, \frac{dy}{dx} \right), \tag{6.76}
$$

$$\varphi_1\left(y(0), \frac{dy}{dx}(0)\right)=0, \tag{6.77}$$

$$\varphi_2\left(y(l), \frac{dy}{dx}(l)\right)=0, \tag{6.78}$$

where f, φ_1, φ_2 are given functions of their variables. Suppose the functions f, φ_1, φ_2 are sufficiently smooth, so that a solution of the boundary value problem (6.76)–(6.78) exists. Suppose that there exists a unique solution of the initial value problem for equation (6.76) for arbitrary initial data, given for $x=0$ or $x=l$. Let us choose some initial conditions $\bar{y}(0)=\bar{y}_0, \frac{d\bar{y}}{dx}(0)=\bar{y}_1$ satisfying the left boundary conditions (6.77). This can be done, for example, by choosing the value $\bar{y}_0$ and solving the equation obtained

$$\varphi_1\left(\bar{y}_0, \frac{d\bar{y}}{dx}(0)\right)=0 \tag{6.79}$$

with respect to $\frac{d\bar{y}}{dx}(0)=y_1$. In view of our assumption, the solution $\bar{y}(x)$ of equation (6.76) satisfying the chosen initial condition is unique. If the function $\bar{y}(x)$ which we thus obtain also satisfies the right-hand side boundary conditions

$$\varphi_2\left(\bar{y}(l), \frac{d\bar{y}}{dx}(l)\right)=0, \tag{6.80}$$

then the given boundary value problem will be solved. However, in the general case, the function $\bar{y}(x)$ which we obtain does not satisfy the right-hand side boundary condition (6.80). By our method of construction, the function $\bar{y}(x_0)$ depends on the value y_0 as on a parameter: $\bar{y}=\bar{y}(x, \bar{y}_0)$, where, in view of general properties of the initial value problem's solution studied in Chapter 2, this dependence is continuous. Changing the value of $\bar{y}_0$ and repeating the algorithm described above, we shall obtain a one-parameter family $\bar{y}(x, \bar{y}_0)$ of solutions of the initial value problems for equation (6.76) which satisfy the left-hand side boundary condition (6.77). The problem is to choose a function $y(x)$ from this family which also satisfies right-hand side boundary condition (6.78). This can be achieved in different ways. For example, we can choose the values $\bar{y}_0$ at random until for some close values of $\bar{y}_{0,i-1}$ and $\bar{y}_{0,i}$ we obtain left-hand sides in (6.80) which have opposite signs. This means that the required value of $\bar{y}_0$ has been "caught in a fork" and we can continue our "shooting" to pinpoint the target, i.e. to specify the required value of $\bar{y}_0$. Since, as we pointed out above, the solutions of the boundary value problems depend continuously on the initial data, the required value of $\bar{y}_0$ lies between $\bar{y}_{0,i-1}$ and $\bar{y}_{0,i}$. This method for solving the boundary value problem (6.76)–(6.78) is known as the shooting method.

If the solution of the Cauchy problem $\bar{y}(x, \bar{y}_0)$ can be found in quadratures, then relation (6.80) is in general a transcendental equation

$$\psi(y_0) = \varphi_2\left(\bar{y}(l, y_0), \frac{d\bar{y}}{dx}(l, y_0)\right) = 0 \qquad (6.81)$$

with respect to the required value y_0. Thus the problem has been reduced to the algebraic problem of finding a root of the equation (6.81). Having solved this equation (analytically or numerically), we obtain the initial conditions $y(0), \frac{dy}{dx}(0)$ which are satisfied by the solution of the given boundary value problem.

In the general case, the functions $\bar{y}(x_0, \bar{y}_0)$ do not have an explicit analytic expression and can only be found numerically for each specific value of the parameter $\bar{y}_0$. Then we can use numerical methods for solving equation (6.81) – say different modifications of Newton's method, the parabola method, etc. For example, for values of the parameter $\bar{y}_0$ equal to $\bar{y}_{0,i-1}$ and $\bar{y}_{0,i}$, we can use the above algorithm to construct the functions $\bar{y}(x, \bar{y}_{0,i-1})$ and $\bar{y}(x, \bar{y}_{0,i})$. Then the new value of the parameter $\bar{y}_{0,i+1}$, according to the method of secants, can be found from the relation

$$\bar{y}_{0,i+1} = \bar{y}_{0,i} - \frac{(\bar{y}_{0,i} - \bar{y}_{0,i-1})\psi(\bar{y}_{0,i})}{\psi(\bar{y}_{0,i}) - \psi(\bar{y}_{0,i-1})}. \qquad (6.82)$$

Note that, the iteration process (6.82) converges very rapidly if the initial value $\bar{y}_0$ is chosen near the root of equation (6.81).

The above considerations are valid in the general case of a non-linear boundary value problem. In the case of the linear boundary value problem

$$L[y] = \frac{d}{dx}\left[p(x)\frac{dy}{dx}\right] - q(x)y(x) = f(x), \qquad (6.83)$$

$$\alpha_1 y'(0) + \beta_1 y(0) = 0, \qquad (6.84)$$

$$\alpha_2 y'(l) + \beta_2 y(l) = 0, \qquad (6.85)$$

it is possible to propose a modification of the shooting method which yields a solution of the given boundary value problem by reducing it to the solution of a relatively small number of initial-value problems.

Construct the function $y_1(x)$ which is the solution of the problem for the non-homogeneous equation (6.83) with initial conditions

$$y_1(0) = \alpha_1, \quad y_1'(0) = -\beta_1. \qquad (6.86)$$

Obviously, in view of (6.86), the function $y_1(x)$ satisfies the left-hand side boundary condition (6.84). Determine the function $y_0(x)$ as the solution of the problem for the homogeneous equation (6.83) with the same initial conditions

$$y_0(0) = \alpha_1, \quad y_0'(0) = -\beta_1. \tag{6.86'}$$

In view of the linearity of the problem and of the homogeneous conditions, the function $C y_0(x)$ is also a solution of the homogeneous equation (6.83) and satisfies the left-hand side boundary condition (6.84). Therefore, the solution of the given boundary value problem may be constructed in the form

$$y(x) = y_1(x) + C y_0(x). \tag{6.87}$$

The constant C may be determined from the fact that the function (6.87) must satisfy the right-hand side boundary condition (6.85):

$$\alpha_2 y_1'(l) + \beta_2 y_1(l) + C[\alpha_2 y_0'(l) + \beta_2 y_0(l)] = 0. \tag{6.88}$$

Note that by the assumption concerning the uniqueness of the solution of the given boundary value problem, the square bracket in (6.88) is necessarily non-zero, and therefore the constant C in this relation is uniquely determined. Thus, in the case considered, the solution of the boundary value problem reduces to the solution of only two initial value problems, which may be carried out by the numerical methods proposed in section 1 of the present chapter.

Remarks. 1. The solution of boundary value problems for equations and systems of orders higher than the second may also be obtained by the shooting method. The main ideas of the method need not be essentially changed; however, instead of one equation (6.81), in the general case we shall obtain p transcendental equations of the (6.81) type with respect to the parameters $y_1, y_2, \ldots, y_p$, where p is the number of left-hand side boundary conditions. This leads to considerable technical difficulties in the practical solution of high order systems by the shooting method. Notable exceptions are linear systems, for which the problem reduces to the solution of a system of linear algebraic equations.

2. The shooting method reduces the solution of the given boundary value problem to that of a series of auxiliary initial value problems. Estimates of the error in the numerical solution of the initial value problem in terms of the errors in the initial data and in the computation of the right-hand sides of the equations (estimate (6.43) and others) obtained above (§1) shows that this error may increase exponentially with the increase of the length of the segment $[0, l]$ on which the boundary value problem is solved. Thus it becomes necessary either to devise an important modification of the shooting method (such is the so-called method of differential orthogonal factorisation [1]), or to apply other methods to the solution of boundary value problems, e.g. direct finite difference methods, which we shall study in the next subsection.

2. Finite difference methods. As we have already pointed out, the essence of these methods consists in replacing the given problem (for a differential equation) by a system of algebraic equations (for the values of the net function) which approximates the solution of the given problem on the net. Let us consider the main ideas of this method in the case of the simplest linear boundary value problem for second order equations

$$y'' - q(x)y = f(x) \tag{6.89}$$

$$y(0) = y(l) = 0. \tag{6.90}$$

It is easy to see that the boundary value problem (6.89), (6.90) has a unique solution when $q(x) > 0$. This statement is a corollary of the general property of eigenvalue problems obtained in Chapter 4. As was shown, when $q(x) > 0$, all the eigenvalues λ_n of the corresponding eigenvalue problem are strictly positive: $\lambda_n > 0$. Since $\lambda = 0$ is not an eigenvalue of this problem, the boundary value problem (6.89), (6.90) has a unique solution.

The uniqueness of the solution of problem (6.89), (6.90) may be proved directly by using the so-called maximum principle. Suppose that there exists a nontrivial solution of the homogeneous boundary value problem (6.89), (6.90) (the function $y_0(x)$). This function is continuous on the closed interval $[0, l]$ and therefore, by the well-known property of continuous functions, it assumes its maximal value M at some point x_0 of this interval:

$$y_0(x) \le M = y_0(x_0) \qquad \forall x \in [0, l] \tag{6.91}$$

where $M \ge 0$, since $y_0(0) = y_0(l) = 0$. The point x_0 may either be a boundary point or an inner point of the segment $[0, l]$. Suppose x_0 is a boundary point. Then $M = 0$, since $y_0(0) = y_0(l) = 0$. Now assume x_0 is an inner point; then we have a maximum at this point and $y''(x_0) \le 0$. But it follows from the equation itself (see (6.89)), for $f(x) = 0$, that

$$0 \ge y_0''(x_0) = q(x_0)y_0(x_0) = q(x_0)M \ge 0, \tag{6.92}$$

and this is possible only if $M = 0$. Thus $M = 0$. In a similar way, it can be shown that the minimal value m of the function $y_0(x)$ on the closed interval $[0, l]$ equals zero: $m = 0$. This implies $y_0(x) \equiv 0$, $x \in [0, l]$, which proves the statement about the uniqueness of the solution of the boundary value problem (6.89), (6.90).

Now let us turn to the construction of the difference scheme which approximates the problem (6.89), (6.90). On the segment $[0, l]$ introduce a net ω_h and consider the net function $y_{(h)}$ defined at the nodes of the net. Approximate the second derivative $y''(x)$ by the difference quotient

$$\frac{y_{i+1} - 2y_i + y_{i-1}}{h^2} \tag{6.93}$$

and replace the original boundary value problem by the finite difference scheme

$$L_h y_{(h)} = \begin{cases} \dfrac{y_{(h)i+1} - 2y_{(h)i} + y_{(h)i-1}}{h^2} - q_{(h)i} y_{(h)i} \\ y_{(h)0} \\ y_{(h)n} \end{cases} = \begin{Bmatrix} f_{(h)i} \\ 0 \\ 0 \end{Bmatrix}, \qquad (6.94)$$

where $q_{(h)i} = q(x_i)$, $f_{(h)i} = f(x_i)$.

The scheme (6.94) is obviously a system of $n+1$ linear algebraic equations with respect to the values of the net function at $n+1$ nodes of the net. In studying this scheme, we must answer questions concerning the existence of the solution, its stability and the convergence of the family of net functions to the solution of the given problem when $h \to 0$ and indicate an algorithm for constructing the net functions. We first discuss the existence of solutions for the scheme (6.94). To show the existence and uniqueness of the solution for the non-homogeneous system, it is sufficient to establish the absence of nontrivial solutions for the corresponding homogeneous system, since the scheme (6.94) is a system of linear equations. Assume that the homogeneous system has a nontrivial solution $y_{(h)0} = 0$, $y_{(h)1}, \ldots, y_{(h)n} = 0$. Denote the largest of these $n+1$ numbers by M:

$$y_{(h)i} \le M = y_{(h)k}. \qquad (6.95)$$

Since $y_{(h)0} = y_{(h)n} = 0$, we have $M \ge 0$. We shall now argue as in the proof of the non-existence of non-trivial solutions for the homogeneous equation (6.89). Suppose the value M is assumed at one of the boundary nodes. Then $M = 0$. If M is achieved at an interior node $i = k$, then, by the homogeneous equation (6.94), we have

$$(2 + h^2 \cdot q_{(h)k}) y_{(h)k} = (2 + h^2 \cdot q_{(h)k}) M = y_{(h)k+1} + y_{(h)k-1} \le 2M. \qquad (6.96)$$

If $M > 0$, then this implies a contradiction, since by condition $q(x) > 0$ we have $(2 + h^2 \cdot q_{(h)k}) M > 2M$. Therefore in this case we also have $M = 0$. Thus $M = 0$. In a similar way it can be shown that the minimal value m of the $n+1$ numbers $y_{(h)0}, y_{(h)1}, \ldots, y_{(h)n}$ also equals zero: $m = 0$. Therefore all the values $y_{(h)0}, y_{(h)1}, \ldots, y_{(h)n}$ vanish, which means that there are no non-trivial solutions of the homogeneous system of linear algebraic equations (6.94). Therefore, the difference scheme (6.94) has a unique solution for any right-hand side $f_{(h)}$ for any n.

Now let us study the approximation properties of the scheme (6.94). Suppose the functions $q(x)$ and $f(x)$ have continuous second derivatives on the segment $[0, l]$. Then it follows from (6.89) that the solution of the boundary value problem (6.98), (6.90) possesses continuous and bounded derivatives up to the fourth order on the segment $[0, l]$. Expanding the exact solution of the boundary value problem into a Taylor series and substituting this expansion into the left-hand side of equation (6.94), we get

$$
\frac{1}{h^2} \left\{ y(x_i) + hy'(x_i) + \frac{h^2}{2} y''(x_i) + \frac{h^3}{6} y'''(x_i) \right.
$$

$$
+ \frac{h^4}{24} y^{(IV)}(x_i + \theta_1 h) - 2y(x_i) + y(x_i) - hy'(x_i) + \frac{h^2}{2} y''(x_i)
$$

$$
\left. - \frac{h^3}{6} y'''(x_i) + \frac{h^4}{24} y^{(IV)}(x_i - \theta_2 h) \right\} - q(x_i) y(x_i). \tag{6.97}
$$

$$
(0 \le \theta_1 \le 1, \quad 0 \le \theta_2 \le 1).
$$

therefore

$$
y''(x_i) - q(x_i) y(x_i) + \frac{h^2}{24} [y^{(IV)}(x_i + \theta_1 h) + y^{(IV)}(x_i - \theta_2 h)] = f(x_i), \tag{6.98}
$$

which implies that the order of approximation of scheme (6.94) equals 2.

Now let us study the stability of the difference scheme (6.94) with respect to a perturbation of the boundary conditions and of the right-hand side of the equation. The perturbed problem will be of the form

$$
L_h \bar{y}_{(h)} = \left\{ \begin{array}{c} \dfrac{\bar{y}_{(h)i+1} - 2\bar{y}_{(h)i} + \bar{y}_{(h)i-1}}{h^2} - q_{(h)i} \bar{y}_{(h)i} \\ \bar{y}_{(h)0} \\ \bar{y}_{(h)n} \end{array} \right\} = \left\{ \begin{array}{c} f_{(h)i} + \delta f_{(h)i} \\ \varepsilon_0 \\ \varepsilon_n \end{array} \right\}. \tag{6.99}
$$

Denote the error of the solution by

$$
\delta y_{(h)i} = \bar{y}_{(h)i} - y_{(h)i} \tag{6.100}
$$

and subtract (6.99) from (6.94); then, by linearity,

$$
(2 + h^2 \cdot q_{(h)i}) \delta y_{(h)i} = \delta y_{(h)i+1} + \delta y_{(h)i-1} - h^2 \delta f_{(h)i},
$$

$$
\delta y_{(h)0} = \varepsilon_0, \quad \delta y_{(h)n} = \varepsilon_n. \tag{6.101}
$$

Choose the value $i = k$ $(i = 1, \ldots, n - 1)$ for which the absolute value of the error is the greatest possible

$$
|\delta y_{(h)k}| \ge |\delta y_{(h)i}| \quad (i = 1, \ldots, n - 1). \tag{6.102}
$$

It follows from equation (6.101) that

$$
(2 + h^2 q_{(h)k}) |\delta y_{(h)k}| \le |\delta y_{(h)k+1}| + |\delta y_{(h)k-1}| + h^2 |\delta f_{(h)k}|. \tag{6.103}
$$

The inequality (6.103) will only be strengthened if we replace $|\delta y_{(h)k+1}|$ and $|\delta y_{(h)k-1}|$ by the maximal value $|\delta y_{(h)k}|$, while $|\delta f_{(h)k}|$ is replaced by the maximal value $\|\delta f_{(h)}\|$. Then (6.103) yields

$$h^2 q_{(h)k}|\delta y_{(h)k}| \le h^2 \|\delta f_{(h)}\|. \tag{6.104}$$

We therefore obtain

$$|\delta y_{(h)k}| \le \frac{\|\delta f_{(h)}\|}{\min_{[0,l]} q(x)} \quad (k=1,\dots,n-1),$$

but since $|\delta y_{(h)k}| \le \max \{|\varepsilon_0|, |\varepsilon_n|\}\ (k=0,n)$, we finally get

$$\|\delta y_{(h)}\| \le \max \left\{ \frac{\|\delta f_{(h)}\|}{\min_{[0,l]} q(x)}, |\varepsilon_0|, |\varepsilon_n| \right\} \le C \|\delta \varphi_{(h)}\|_1, \tag{6.105}$$

which proves the stability of scheme (6.94) with respect to the boundary conditions and the right-hand side of the equation.

The approximation and stability properties of scheme (6.94) established above enable us, by using Theorem (6.1), to claim the validity of the following theorem.

Theorem 6.3. *If the functions $q(x)>0$ and $f(x)$ possess continuous second derivatives on the segment $[0, l]$, then the family $y_{(h)}$ of solutions of the difference schemes (6.94) converges, as $h \to 0$, to an exact solution of problem (6.89), (6.90) and the order of convergence is equal to 2.*

Now let us discuss how scheme (6.94) can actually be carried out. Unlike the previous difference schemes for the solution of initial value problems (the Euler and Runge-Kutta schemes), the given scheme does not give an explicit algorithm for the successive computation of the values of the net functions at the nodes of the net, but is a system of linear algebraic equations in which the unknowns are the values of the net function at all the nodes of the net. Therefore, for the numerical solution of problem (6.94), one may use general methods for solving linear algebraic systems. However, in case of a sufficiently high order n, these methods turn out to be quite laborious. It is natural to make use of the special properties of the matrix of the algebraic system obtained previously. In the given case, the matrix is three-diagonal: only the main diagonal and the two nearest smaller diagonals have nonzero elements. Thus it is possible to propose an extremely effective special method for the solution of system (6.94); we now pass to its exposition.

3. The sweep method (successive substitution) of the solution of algebraic systems with three-diagonal matrix. We will consider this method for the following system, which is a generalization of scheme (6.94):

$$A_i y_{i-1} - C_i y_i + B_i y_{i+1} = -F_i \quad (i=1,\ldots,n-1), \qquad (6.106)$$

$$y_0 = \alpha y_1 + \beta, \quad y_n = \gamma y_{n-1} + \delta. \qquad (6.107)$$

Suppose the coefficients which are involved in equation (6.106) and in the boundary conditions (6.107) satisfy the assumptions

$$A_i, B_i, C_i > 0; \quad C_i \geq A_i + B_i;$$

$$0 \leq \alpha < 1; \quad 0 \leq \gamma < 1. \qquad (6.108)$$

It is obvious that scheme (6.94) considered above satisfies conditions (6.108). Thus, as in the case of scheme (6.94), it is not difficult to show that the problem can be solved and solved uniquely under assumption (6.108).

We shall attempt to determine coefficients α_i, β_i such that for all values of the index $i = 1, 2, \ldots, n$ the following relation

$$y_{i-1} = \alpha_i y_i + \beta_i \qquad (6.109)$$

holds. Substituting the required form of solution (6.109) in equation (6.106), we obtain

$$(A_i \alpha_i - C_i) y_i + B_i y_{i+1} + (A_i \beta_i + F_i) = 0. \qquad (6.110)$$

Expressing y_i in terms of y_{i+1} by formula (6.109), we can rewrite (6.110) in the form

$$[(A_i \alpha_i - C_i)\alpha_{i+1} + B_i] y_{i+1} + [(A_i \alpha_i - C_i)\beta_{i+1} + A_i \beta_i + F_i] = 0. \quad (6.111)$$

The relation (6.111) will be identically satisfied for $i = 1, 2, \ldots, n-1$ if we require that each of the square brackets vanish for $i = 1, 2, \ldots, n-1$. This requirement gives recurrent relations for successively determining the coefficients α_{i+1} and β_{i+1} from the given values $\alpha = \alpha_1$ and $\beta = \beta_1$, namely

$$\alpha_{i+1} = \frac{B_i}{C_i - A_i \alpha_i}, \quad \beta_{i+1} = \frac{A_i \beta_i + F_i}{C_i - A_i \alpha_i}. \qquad (6.112)$$

It is easy to show that if conditions (6.108) are met, then the denominators in formulas (6.112) do not vanish and, moreover, $0 \leq \alpha_i < 1$. Indeed, if we rewrite the second condition (6.108) in the form

$$C_i = A_i + B_i + D_i, \quad D_i \geq 0, \qquad (6.113)$$

then the first formula (6.112) may be written as

$$\alpha_{i+1} = \frac{B_i}{B_i + A_i(1-\alpha_i) + D_i}. \qquad (6.114)$$

Since $0 \le \alpha_1 < 1$, $A_i, B_i > 0$ and $D_i \ge 0$, it follows that, for all the α_i computed according to formula (6.114), we also have the condition $0 \le \alpha_i < 1$. Thus all the denominators in formula (6.112) are strictly greater than zero, which enables us to find all the coefficients α_i, β_i $(i = 1, \ldots, n)$.

Having computed α_n and β_n from (6.109), we obtain

$$y_{n-1} = \alpha_n y_n + \beta_n. \qquad (6.115)$$

On the other hand, from the boundary condition (6.107), we have

$$y_n = \gamma y_{n-1} + \delta = \gamma(\alpha_n y_n + \beta_n) + \delta \qquad (6.116)$$

so that

$$y_n = \frac{\gamma \beta_n + \delta}{1 - \gamma \alpha_n}.$$

Previously, we showed that $0 \le \alpha_n < 1$. Therefore, in view of condition (6.108) $(0 \le \gamma < 1)$, the denominator of this expression does not vanish. Therefore y_n is defined. Now, according to formula (6.109), we can successively compute all the unknowns y_{i-1} $(i = n, n-1, \ldots, 1)$ ("backward sweep").

Remarks. 1. By the relation $0 \le \alpha_i < 1$ that we have established, neither in the forward nor in the backward sweep will there be any compilation of errors.

2. We have outlined an application of the sweep method in which we first determine the coefficients by transforming the left boundary condition and then recover the solution by using formula (6.109). Obviously, in a similar way, we can find the coefficients γ_i, δ_i by sweeping from the right-hand side boundary condition from right to left and recover the solution by sweeping back from left to right.

Chapter VII
Asymptotics of Solutions of Differential Equations with Respect to a Small Parameter

The necessity of developing approximate solution methods for differential equations was pointed out above. In Chapter 6, we dealt with the so-called numerical methods of solution of differential equations; these methods give a table of values of the solution at nodes of the net with any chosen precision for any equation with appropriate supplementary conditions. This chapter will be concerned with another class of approximation methods, the so-called asymptotic methods, whose aim is to obtain formulas describing the solution's qualitative behaviour on some interval of values of the independent variable. The precision of such formulas possesses natural limitations (this will be explained in detail below). Note at once that numerical and asymptotic methods do not exclude, but rather supplement each other.

§1. Regular Perturbations

1. The notion of asymptotic representation. In §5 of Chapter 2, we studied in detail the dependence of the solution of the initial value problem on the parameters involved in the equation. In the present section, we put $t_0 = 0$, which can always be achieved by a change of the independent variable and limit ourselves to the scalar case (y is a scalar, μ is a scalar) which is not an essential assumption, but is made to simplify the exposition. Thus we consider the initial value problem

$$\frac{dy}{dt} = f(y, t, \mu) \qquad y(0, \mu) = y^0. \tag{7.1}$$

Here we shall assume that μ varies in a neighbourhood of the value $\mu = 0$; this may also be achieved by an appropriate choice of the origin on the μ-axis.

In Chapter 2 (Theorem 2.10), it was proved that, for appropriate conditions on the right-hand side of (7.1), the solution $y(t, \mu)$ of problem (7.1) exists and is a continuous function of t and μ on the set $t \in [0, T]$, $|\mu| < c$.

This theorem gives us the following possibility of constructing the approximate solution of problem (7.1). Consider the problem obtained from (7.1) by formally putting $\mu = 0$ in it:

$$\frac{dy}{dt} = f(y, t, 0), \quad y(0) = y^0. \tag{7.2}$$

The problem (7.2) is, in general, simpler than the original problem (7.1) and its solution (which we shall denote by $\bar{y}(t)$) is easier to study and perhaps it can even be effectively constructed. It follows from Theorem 2.11 that, on the segment $[0, T]$, we have

$$y(t, \mu) = \bar{y}(t) + \varepsilon(t, \mu) \tag{7.3}$$

where $\varepsilon(t, \mu) \to 0$ uniformly as $\mu \to 0$. Thus $\bar{y}(t)$ is an approximation of $y(t, \mu)$, while $\varepsilon(t, \mu)$ is the error of this approximation.

Formula (7.3) is the simplest version of the so-called *asymptotic formula* (or *asymptotic representation*) of the solution $y(t, \mu)$ with respect to the small parameter μ. By asymptotic formulas with respect to a small parameter, we mean formulas in which certain terms, called remainders or residual terms, are not written out explicitly – only their properties for $\mu \to 0$, e.g. the order with which they tend to zero as $\mu \to 0$, are indicated.

In formula (7.3), $\varepsilon(t, \mu)$ is the remainder. Theorem 2.11 enables us to indicate the order with which $\varepsilon(t, \mu)$ tends to zero. Indeed, the existence of the derivative with respect to μ allows us to write

$$y(t, \mu) = \bar{y}(t) + \frac{\partial y}{\partial \mu}(t, \theta\mu)\mu \quad (0 \le \theta \le 1) \tag{7.4}$$

thus establishing that $\varepsilon(t, \mu) = O(\mu)$. Here and in the sequel equalities of this type indicate that $|\varepsilon(t, \mu)| \le C\mu$, where C is a certain constant which does not depend on μ for sufficiently small μ.

The smaller the value of μ, the better $\bar{y}(t)$ approximates $y(t, \mu)$. However, in real problems, μ is a small but not an infinitely small expression. Hence an asymptotic formula does not guarantee an arbitrary degree of precision and this is its most important defect. Nevertheless, asymptotic formulas are very convenient in cases when it is necessary to obtain a qualitative picture of the solution.

Remark. The above concerns problems in which the small parameter is a physically natural small parameter. There exist, however, problems of a different type, e.g. problems arising in establishing computational algorithms, when a certain small parameter whose value can be changed at will is introduced artificially (for example the size of the step). In problems of this type an arbitrary degree of precision can be achieved.

The results of Chapter 2 give an asymptotic formula for $y(t, \mu)$ with a remainder tending to zero with degree higher than $O(\mu)$, whenever $f(y, t, \mu)$ satisfies conditions guaranteeing the existence of enough continuous derivatives with respect to μ. Let us state this consequence of §5, Chapter 2, in the form of a separate theorem.

Theorem 7.1. *Suppose that in some domain D of the variables y, t, μ, the function $f(y, t, \mu)$ possesses continuous and uniformly bounded partial derivatives with respect to y and μ to the order $n+1$ inclusive. Then there exists a closed interval $[0, T]$ on which for the solution $y(t, \mu)$ of problem (7.1) we have the asymptotic representation*

$$y(t, \mu) = \bar{y}(t) + \mu \frac{\partial y}{\partial \mu}(t, 0) + \ldots + \frac{\mu^n}{n!} \frac{\partial^n y}{\partial \mu^n}(t, 0) + \varepsilon_{n+1}(t, \mu), \qquad (7.5)$$

where $\varepsilon_{n+1}(t, \mu) \to 0$ uniformly as $\mu \to 0$, $0 \le t \le T$, and $\varepsilon_{n+1}(t, \mu) = O(\mu^{n+1})$.

Remarks. 1. The expressions $\dfrac{\partial^k y}{\partial \mu^k}(t, 0)$ can be determined from the equations of variations written out in §5, Chapter 2, The representation (7.5) may be obtained in another way. Let us substitute y, written as a formal series, into (7.1):

$$y = y_0(t) + \mu y_1(t) + \ldots . \qquad (7.6)$$

If after the substitution we also expand the expression $f(y_0 + \mu y_1 + \ldots, t, \mu)$ into a formal power series, we get

$$\frac{dy_0}{dt} + \mu \frac{dy_1}{dt} + \ldots = f(y_0, t, 0) + \frac{\partial f}{\partial y}(y_0, t, 0)(\mu y_1 + \ldots) + \frac{\partial f}{\partial \mu}(y_0, t, 0)\mu$$

$$+ \ldots + \frac{1}{n!}\left((\mu y_1 + \ldots)\frac{\partial}{\partial y} + \mu \frac{\partial}{\partial \mu}\right)^n f(y_0, t, 0) + \ldots; \qquad (7.7)$$

$$y_0(0) + \mu y_1(0) + \ldots = y^0.$$

Putting terms in the same powers of μ equal to each other, we get

$$\frac{dy_0}{dt} = f(y_0, t, 0), \qquad y_0(0) = y^0, \qquad (7.7')$$

$$\frac{dy_1}{dt} = \frac{\partial f}{\partial y}(y_0, t, 0)y_1 + \frac{\partial f}{\partial \mu}(y_0, t, 0), \qquad y_1(0) = 0. \qquad (7.7'')$$

. . . .

Solving these problems successively, let us determine the terms of the series (7.6). Problem (7.7') for $y_0(t)$ coincides with problem (7.2) and therefore, by uniqueness, we have $y_0(t) = \bar{y}(t)$.

Problem (7.7″) coincides with the problem for $\dfrac{\partial y}{\partial \mu}(t,0)$ (see 2.149) and therefore $y_1(t) = \dfrac{\partial y}{\partial \mu}(t,0)$, etc.

2. If $f(y,t,\mu)$ possesses continuous and uniformly bounded derivatives of any order in D, then the value of n in (7.5) is arbitrary. Thus for $y(t,\mu)$ the Maclaurin series with general term $\mu^n y_n(t) = \mu^n \dfrac{\partial^n y}{\partial \mu^n}(t,0)$ is defined and is such that the difference $\varepsilon_{n+1}(t,\mu)$ between the partial sum of the series and the solution $y(t,\mu)$ is $O(\mu^{n+1})$. Such a series is known as an asymptotic series, or an asymptotic expansion with respect to the small parameter μ for $y(t,\mu)$. Let us stress that $\varepsilon_{n+1}(t,\mu) = O(\mu^{n+1})$ for fixed n as $\mu \to 0$. In the case when μ is fixed while $n \to \infty$, it is possible that $\varepsilon_{n+1}(t,\mu)$ does not have a limit, i.e. the series thus constructed is not necessarily convergent.

3. Instead of the expressions "asymptotic formula", "asymptotic representation", "asymptotic expansion", the briefer term "asymptotics" is often used.

4. The theory developed in §5, Chapter 2, and theorem 7.1 which follows from it, give a mathematical basis for neglecting the small terms of an equation, as is often done in physics. These small terms are usually called *perturbations*. In this connection, equation (7.2) is often called the *unperturbed equation*, while equation (7.1) is the *perturbed equation*. The theory dealing with the foundations of asymptotics with respect to a small parameter is often referred to as the theory of perturbations.

Theorem 7.1 remains valid under conditions of sufficient smoothness (or, as one says, of regularity) of the right-hand side of (7.1) with respect to y and μ. Perturbations which satisfy the requirements of Theorem 7.1 are called *regular perturbations*. This explains the title of the present section.

Example. Let us obtain an asymptotic formula with remainder $O(\mu^2)$, valid on $[0, T]$, for the solution $y(t,\mu)$ of the problem

$$\frac{dy}{dt} = a(t)y + b(t) + \mu c(t) y^2, \quad y(0) = 0.$$

This is the Riccati equation, whose solution cannot be effectively obtained. The functions $y_0(t)$ and $y_1(t)$ can be obtained in quadratures, namely

$$\frac{dy_0}{dt} = a(t)y_0 + b(t), \quad y_0(0) = 0,$$

and therefore

$$y_0(t) = \int_0^t b(\tau) e^{\int_\tau^t a(t)dt} d\tau; \quad \frac{dy_1}{dt} = a(t)y_1 + c(t) y_0^2, \quad y_1(0) = 0,$$

so that

$$y_1(t) = \int_0^t c(\tau) y_0^2(\tau) e^{\int_\tau^t a(t)dt} d\tau; \qquad y(t,\mu) = y_0(t) + \mu y_1(t) + O(\mu^2).$$

2. The existence of the perturbed problem's solution. The results obtained in §5, Chapter 2, have the following special trait: the validity of the asymptotic representation is guaranteed on some segment $[0, T]$, determined by the properties of the right-hand side of (7.1), simultaneously with the existence and uniqueness both of the unperturbed and of the perturbed equations.

The problem may be stated in another way. Assume that the solution of the unperturbed equation (7.2) exists, is unique and belongs to some domain G in the space of the variables (y, t) when $0 \leq t \leq T$. The value of T in this case may be established directly from the explicit form of $\bar{y}(t)$. Will the solution of problem (7.1) also exist on the entire segment $[0, T]$ for a sufficiently small μ and satisfy formula (7.3)? This question is answered by the following

Theorem 7.2. *Suppose that in the domain $G = \{0 \leq t \leq T, |y| < b, |\mu| < \bar{\mu}\}$ the function $f(y, t, \mu)$ is continuous with respect to the set of its variables and satisfies the Lipschitz condition*

$$|f(y_1, t, \mu) - f(y_2, t, \mu)| \leq N|y_1 - y_2|,$$

where N is the same constant for all μ on the segment $|\mu| \leq \bar{\mu}$. Suppose the solution $\bar{y}(t)$ of problem (7.2) exists, is unique on $[0, T]$ and belongs to $D = \{0 \leq t \leq T, |y| < b\}$. Then for every sufficiently small μ the solution $y(t, \mu)$ of problem (7.1) also exists and is unique on $[0, T]$, belongs to D, and we have the following limit uniformly with respect to t

$$\lim_{\mu \to 0} y(t, \mu) = \bar{y}(t). \tag{7.8}$$

Proof. In 7.1 let us turn to a new unknown function $\Delta = y - \bar{y}(t)$, which is the solution of the initial value problem (compare with (2.137))

$$\frac{d\Delta}{dt} = [f(\bar{y} + \Delta, t, \mu) - f(\bar{y}, t, \mu)] + [f(\bar{y}, t, \mu) - f(\bar{y}, t, 0)], \quad \Delta(0) = 0. \tag{7.9}$$

Consider the following domain of the variables Δ, t: $\tilde{D} = \{0 \leq t \leq T, |\Delta| < C\}$, where $C = b - \beta$, $\beta = \sup_{[0,T]} |\bar{y}(t)|$. For $|\Delta| < C$, we have $|y| < |\bar{y}| + |\Delta| < \beta + C = b$, i.e. for $|\Delta| < C$ the variables of $f(y, t, \mu)$ remain in the domain D. Then, according to the Lipschitz condition, we have

$$|f(\bar{y} + \Delta, t, \mu) - f(\bar{y}, t, \mu)| \leq N|\Delta|,$$

and since $f(y, t, \mu)$ is continuous we conclude that $f(\bar{y}, t, \mu) - f(\bar{y}, t, 0)$ tends to zero uniformly with respect to t as $\mu \to 0$, i.e. there exists a certain function $\omega(\mu)$ depending only on μ which tends to zero as $\mu \to 0$ and satisfies

$$|f(\bar{y}, t, \mu) - f(\bar{y}, t, 0)| < \omega(\mu).$$

Now let us use the same considerations which were applied to prove Theorem 3.2. The existence and uniqueness theorem guarantees the existence and uniqueness of $\Delta(t, \mu)$ and the inequality $|\Delta| < C$ on a certain segment $[0, H]$. Taking $(H, \Delta(H, \mu))$ to be the new initial point, we can extend the solution to a larger segment $[0, H_1]$ $(H_1 > H)$ etc. Suppose $[0, \bar{H})$ $(\bar{H} \le T)$ is the maximal semi-interval on which there exists the unique solution $\Delta(t, \mu)$ of the problem (7.9) belonging to $\tilde{D}$, while $H_n \to \bar{H}$ is an arbitrary sequence. Let us prove that the following limit exists: $\lim\limits_{n \to \infty} \Delta(H_n, \mu)$.

Since for any n we have $|\Delta(t, \mu)| < C$ on $[0, H_n]$, it follows that

$$\left|\frac{d\Delta}{dt}\right| \le N|\Delta| + \omega(\mu);$$

hence, according to Lemma 2.1,

$$|\Delta| \le \frac{\omega(\mu)}{N} (e^{N\bar{H}} - 1) = \omega_1(\mu), \qquad (7.10)$$

where $\omega_1(\mu)$ possesses the same properties as $\omega(\mu)$. Therefore

$$\left|\frac{d\Delta}{dt}\right| \le N\omega_1(\mu) + \omega(\mu).$$

On the basis of this inequality, we can prove the existence of the limit $\lim\limits_{n \to \infty} \Delta(H_n, \mu)$ exactly in the same way as in Theorem 3.2. Since $|\Delta(H_n, \mu)| \le \omega_1(\mu) < C$, we also have $\lim\limits_{n \to \infty} \Delta(H_n, \mu) \le \omega_1(\mu) < C$. Further arguments coincide word for word with those which were used in Theorem 3.2 and lead us to the conclusion that $\bar{H} = T$ while $\Delta(t, \mu)$ exists and is unique on the segment $[0, T]$. The inequality (7.10) implies the uniform convergence of $\Delta(t, \mu)$ to zero as $\mu \to 0$ i.e. we have (7.8). The theorem is proved.

Remarks. 1. The theorem was proved for the scalar case, but a similar statement remains valid in the case when y is a vector.

2. Theorem 7.2 remains valid not only if the equation is perturbed but also the initial data i.e. if (7.1) is of the form

$$\frac{dy}{dt} = f(y, t, \mu), \qquad y(\omega_1(\mu), \mu) = y^0 + \omega_2(\mu).$$

§ 2. Singular Perturbations

In applications, one often meets cases when the small parameter μ is involved in the equation in such a way that the theory of the previous section cannot be applied. Consider, as a simple example, the motion of a pendulum (see Chapter 1, § 2),

$$\mu y'' + \alpha y' + k y = f(t) \tag{7.11}$$

where $I = \mu$ is a small parameter. In the case considered in the previous section, in order to obtain the approximate expression for the solution, it was possible to formally put $\mu = 0$ in the equation and take the solution of the equation simplified in this way. Can we do the same thing in the case of (7.11)?

The motion of the pendulum in (7.11) is determined by its initial position and velocity $y(0) = y_0$, $y'(0) = y_1^0$. Setting $\mu = 0$ in (7.11), we obtain an equation of *lower* (first) order, whose solution is determined only by $y(0)$. Thus it is clear from the outset that, if we do this, we shall not be able to take into consideration all the factors which determine the solution of (7.11) and, at least in the neighbourhood of the initial point, we shall not obtain a correct model.

Thus the conclusions of the previous section are no longer valid in our case, and therefore the assumptions of the theorem discussed in that section do not hold. In order to understand which conditions are no longer met, let us write (7.11) in the form (7.1) (the equation will now be a vector equation, but, as was pointed out above, this is not important):

$$z' = \frac{-\alpha z - k y + f(t)}{\mu} = f_1(z, y, t, \mu)$$

$$y' = z = f_2(z, y, t, \mu)$$

We see now that f_1 is not a continuous function of μ when $\mu = 0$, i.e. the main assumption of the theorem of the previous section – continuity of the right-hand side – no longer holds. In other words, we can say that, in the given case, the right hand side depends on μ in a *non-regular*, or *singular* fashion. Hence, perturbations of the $\mu y''$ type, i.e. such that the small parameter is a factor of the highest derivative, are known in the literature as *singular perturbations*.

Simplest examples show that singular perturbations of the system possess a series of properties which are basically different from the ones of regularly perturbed systems studied in § 1.

Consider the equation

$$\mu y' = a y + b; \quad a = \text{const}, \quad b = \text{const}, \tag{7.12}$$

with initial condition $y(0, \mu) = y^0$. Its exact solution is of the form

$$y(t, \mu) = \left(y^0 + \frac{b}{a} \right) e^{at/\mu} - \frac{b}{a}. \tag{7.13}$$

In equation (7.12) set $\mu = 0$, obtained (in the notation of §1) $\bar{y} = -b/a$. Analysing (7.13), we see that y will be near to $\bar{y}$ only if certain special conditions discussed in §1 are met. Namely, if we consider the solution of the initial value problem to the right of $t = 0$, then $y \to \bar{y}$, when $a < 0$ and $\mu \to +0$ (or $a > 0$ and $\mu \to -0$). In the case when μ tends to zero in an arbitrary way, neither for $a < 0$ nor for $a > 0$ does the solution y have a limit; and it turns out to be unbounded. Moreover, even if we have a limit; and it turns out to be unbounded. Moreover, even if we have the conditions $a < 0$, $\mu \to +0$ (or $a > 0$, $\mu \to -0$) the passage to the limit $y \to \bar{y}$ takes place only for those values of t which are strictly positive, since, for $t = 0$, we have $y(0, \mu) = y^0$, while y^0 is in general not equal to $-b/a$.

All these facts lead us to the conclusion that small terms may be neglected only if certain special conditions are met. These are determined by a special theory, to which we now turn.

1. Equations with small parameter multiplying the highest order derivative. Limit theorem. Consider the system of differential equations of the form

$$\mu \frac{dz}{dt} = F(z, y, t), \quad \frac{dy}{dt} = f(z, y, t), \tag{7.14}$$

where $\mu > 0$ is a small parameter. The following initial value problem is posed

$$z(0, \mu) = z^0, \quad y(0, \mu) = y^0. \tag{7.15}$$

This problem includes, as a particular case, the initial value problem for equation (7.11).

1°. The right-hand sides of (7.14) will be assumed continuous with their partial derivatives with respect to z and y in some domain

$$H = \{(y, t) \in \bar{D} = \{0 \le t \le T, |y| \le b\}, |z| < d\}.$$

Formally putting $\mu = 0$ in (7.14), we obtain a unperturbed system of equations or, as it is called in the theory of singular perturbations, a *degenerate system*, since its order is less than the order of system (7.14):

$$0 = F(z, y, t), \quad \frac{dy}{dt} = f(z, y, t). \tag{7.16}$$

In order to determine the solution of this system, it is necessary, first of all, to solve the first of the equations (7.16), which is a finite (not a differential) equation with respect to z. This equation is non-linear and can therefore possess several solutions. We shall assume that all the solutions (roots) $z = \varphi(y, t)$ of this equation are real and are isolated in $\bar{D}$. It is necessary to choose one of the roots $z = \varphi(y, t)$ and substitute it into the second equation (7.16). The question is to find which of the roots must be chosen in order to guarantee that the solution of system (7.16) which we shall construct is close to the solution $z(t, \mu)$, $y(t, \mu)$ of problem (7.14), (7.15). The rule for choosing the root will be stated further (see $3°$, $5°$). After substituting $z = \varphi(y, t)$ into the second equation (7.16), we obtain a differential equation with respect to y and, in order to determine y uniquely, we must have an initial condition. It is natural to assume that of the two initial conditions (7.15) we should leave only one: the one concerning y. Thus we have come to the problem

$$\frac{d\bar{y}}{dt} = f(\varphi(\bar{y}, t), \bar{y}, t), \quad \bar{y}(0) = y^0, \tag{7.17}$$

where $\varphi(y, t)$ is one of the roots of the equation $F(z, y, t) = 0$.

$2°$. We shall assume the function $\varphi(y, t)$ continuous together with its derivative with respect to y when $(y, t) \in \bar{D}$.

Definition. The root $z = \varphi(y, t)$ is said to be *stable* in the domain $\bar{D}$ if, for all points $(y, t) \in \bar{D}$, we have the inequality

$$\frac{\partial F}{\partial z}(\varphi(y, t), y, t) < 0. \tag{7.18}$$

Remark. Compare this to the requirement $a < 0$ in the example considered above.

$3°$. We shall assume that in (7.17) $\varphi(y, t)$ is a stable root.

$4°$. We shall assume that the solution $\bar{y}(t)$ of problem (7.17) is defined on the segment $0 \le t \le T$ and belongs to $D = \{0 \le t \le T, |y| < b\}$.

Before carrying out our investigation of the problem further, let us consider a particular case, namely

$$\mu \frac{dz}{dt} = F(z), \quad z(0, \mu) = z^0. \tag{7.19}$$

This case, first of all, interesting by its geometric clarity; secondly, the corresponding results will be useful in considering the general case (7.14).

Suppose $z = \varphi$ is a stable root of the equation $F(z) = 0$. In this case φ is a constant, while D is any interval of the real line. The stability condition (7.18) is of the form $\dfrac{dF}{dz}(\varphi) < 0$. Suppose that, besides φ, the equation $F(z) = 0$ also possesses the roots $\varphi_1, \varphi_2, \ldots$ and φ_2, φ_1 are the two nearest roots to φ

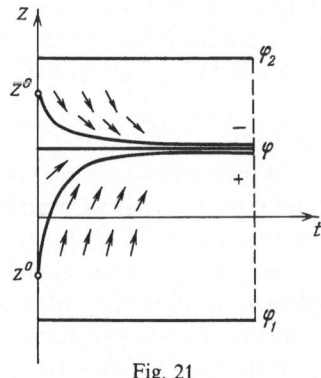

Fig. 21

respectively from above and from below (see Fig. 21, presenting the strip $0 \leq t$ $\leq T$, where T is an arbitrary positive number). Let us study the direction field of equation (7.19). For a sufficiently small μ, the vectors tangent to the integral curves are almost parallel to the z-axis (except in a small neighbourhood of the roots of the equation $F(z) = 0$). The symbols " $+$ " and " $-$ " on the figure indicate the sign of the function $F(z)$. Note that, since $\dfrac{dF}{dz}(\varphi) \neq 0$, the root φ is simple and for $z = \varphi$ the sign of the function $F(z)$ changes.

Suppose $\varphi_1 < z^0 < \varphi_2$. Consider the set of points $|z - \varphi| \leq \varepsilon$ where ε is arbitrarily small (an ε-neighbourhood of the root φ). The character of the direction field leads to the immediate conclusion that an integral curve beginning at the point $(0, z^0)$ (on the figure two possibilities are shown: $z^0 < \varphi$ and $z^0 > \varphi$) will move sharply upward (for $z^0 < \varphi$) or, conversely, downward (for $z^0 > \varphi$) and, having reached an ε-neighbourhood of φ, will no longer leave this neighbourhood as long as μ is sufficiently small. This means that the solution $z(t, \mu)$ of problem (7.19) is close to the solution φ of the degenerate equation $F(z) = 0$ when $\mu \to 0$, if we disregard a certain neighbourhood of $t = 0$ (see the example).

It is also clear from this geometric consideration how important it is for the initial value z^0 to lie in the domain $\varphi_1 < \varphi < \varphi_2$, called the *domain of attraction* (or domain of influence) of the root φ. If, for example, $z^0 < \varphi_1$ and $F < 0$ when $z < \varphi_1$, then the integral curve will move downwards from φ_1 and thus cannot approach φ, while in the case $z^0 < \varphi_1$ and $F > 0$ the curve will approach φ_1, i.e. will not come near φ.

The result which we have obtained from geometric considerations may be stated in the form of the following theorem.

Theorem 7.3. *If φ is a stable root of equation $F(z) = 0$, while the initial condition z^0 is located in its domain of attraction, then the solution $z(t, \mu)$ of problem (7.19) exists on the segment $[0, T]$ and we have the following limiting relation for this solution:*

$$\lim_{\mu \to \infty} z(t, \mu) = \varphi \quad \textit{for} \quad 0 < t \leq T. \tag{7.20}$$

Now let us give an analytic proof of the statement (7.20). Suppose, to be specific, that $z^0 < \varphi$. First of all, consider the domain $z^0 \leq z \leq \varphi - \varepsilon, 0 \leq t \leq \varepsilon$ where $\varepsilon > 0$ is arbitrarily small. In this domain, consider the problem (7.19), taking z as the independent variable $\dfrac{dt}{dz} = \mu \dfrac{1}{F(z)}$, $t(z^0) = 0$. If μ is sufficiently small, then it follows from Theorem 7.2 that there exists a unique solution $t(z)$ of this problem on the segment $[z^0, \varphi - \varepsilon]$ and it is as near as we wish to the vertical $t = 0$ (in the sense of the distance along the t-axis). Moreover, it is positive, since $F(z)$ is positive. Thus the integral curve which beings at the point $(0, z^0)$ enters an ε-neighbourhood of $z = \varphi$ when $t = t_0$ where $t_0 = \omega(\mu) < \varepsilon$.

It is easy to verify that, once it is in an ε-neighbourhood of $z = \varphi$, the integral curve will no longer leave it while $t_0 \leq t \leq T$. To show this, let us use an argument similar to the one used in stability theory (Chapter 5). Introduce the function $V(z) = (z - \varphi)^2$. Assume that the integral curve for $t = t_1 > t_0$ leaves the ε-neighbourhood of the line $z = \varphi$. Then we have $\left.\dfrac{dV}{dt}\right|_{t=t_1} \geq 0$. On the other hand,

$$\left.\frac{dV}{dt}\right|_{t=t_1} = 2[z(t_1, \mu) - \varphi]\frac{F(z(t_1, \mu))}{\mu} < 0$$

by the property $F(z) \lessgtr 0$ for $z \gtrless \varphi$ (which follows from the stability condition $\dfrac{dF}{dz}(\varphi) < 0$). The contradiction thus obtained leads us to conclude that the integral curve remains in an ε-neighbourhood of the line $z = \varphi$. Since it enters this neighbourhood when $t = t_0 < \varepsilon$, it follows, the choice of ε being arbitrary, that (7.20) is valid.

The statement (7.20) may now be rewritten in a different form, convenient for the sequel. Introduce the independent variable $\tau = t/\mu$. Then the problem (7.19) acquires the form

$$\frac{dz}{d\tau} = F(z), \qquad z(0) = z^0. \tag{7.21}$$

The statement (7.20) means that $\lim\limits_{\tau \to \infty} z(\tau) = \varphi$ or, in other words, $\forall \varepsilon \exists \tau_0(\varepsilon)$ such that $\tau \geq \tau_0$ implies the inequality

$$|z(\tau) - \varphi| < \varepsilon. \tag{7.22}$$

Remarks. 1. As we see from the above arguments, the passage to the limit in (7.20) is not uniform with respect to $t \in [0, T]$ which is clearly visible in Fig. 21.

2. It is not by accident that the term "stable root" appears here. The relationship between our constructions and the theory of stability (Chapter 5) is easy to see. Indeed, $z = \varphi$ is an exact solution of equation (7.21) and, by $\dfrac{\partial F}{\partial z}(\varphi)$

< 0, this solution is asymptotically stable in the sense of Lyapunov. In order to verify this, carry out the change of variables $x = z - \varphi$ and use Theorem 5.1 or Theorem 5.3, taking $V = x^2$ for the Lyapunov function. The fact that the integral curve $z = z(\tau)$ remains in the ε-neighbourhood of the line $z = \varphi$ is guaranteed by the stability of the solution $z = \varphi$ of the equation $\dfrac{dz}{d\tau} = F(z)$.

Let us return to the general case (7.14), (7.15). In considering this case, the ideas of stability theory will be applied in conjunction with the ideas of the theory of regular perturbations.

To the problem (7.14), (7.15) assign the problem

$$\mu \frac{dz_0}{dt} = F(z_0, y^0, 0), \qquad z_0(0) = z^0. \tag{7.23}$$

This problem is of the type (7.19), which we have already studied. By $3°$, $z_0 = \varphi(y^0, 0)$ is a stable root of the equation $F(z_0, y^0, 0) = 0$.

$5°$. We shall assume that the initial value z^0 belongs to the domain of attraction of the root $\varphi(y^0, 0)$ of the equation $F(z^0, y^0, 0) = 0$.

Theorem 7.4 *(Tikhonov theorem). If the conditions $1°-5°$ are satisfied, the solution $z(t, \mu), y(t, \mu)$ of problem (7.14), (7.15) exists on $[0, T]$ and we have the limiting relation*

$$\lim_{\mu \to 0} y(t, \mu) = \bar{y}(t) \quad for \quad 0 \le t \le T, \tag{7.24}$$

$$\lim_{\mu \to 0} z(t, \mu) = \varphi(\bar{y}(t), t) \equiv \bar{z}(t) \quad for \quad 0 < t \le T, \tag{7.25}$$

where $\bar{y}(t)$ is the solution of the degenerate problem (7.17).

The proof shall be split up into several steps. 1) Consider at first the neighbourhood of the point $t = 0$. In (7.14) carry out the change of variables $\tau = t/\mu$. Then

$$\frac{dz}{d\tau} = F(z, y, \tau\mu), \qquad z\big|_{\tau=0} = z^0,$$

$$\frac{dy}{d\tau} = \mu f(z, y, \tau\mu), \qquad y\big|_{\tau=0} = y^0. \tag{7.26}$$

On any finite interval of variation of τ, this system may be viewed as a regular perturbed system; the corresponding unperturbed system is of the form

$$\frac{dz_0}{d\tau} = F(z_0, y_0, 0), \qquad z_0\big|_{\tau=0} = z^0,$$

$$\frac{dy_0}{d\tau} = 0, \qquad y_0\big|_{\tau=0} = y^0. \tag{7.27}$$

Hence $y_0(\tau) \equiv y^0$, while

$$\frac{dz_0}{d\tau} = F(z_0, y^0, 0), \quad z_0|_{\tau=0} = z^0. \tag{7.28}$$

This problem is equivalent to problem (7.23) and is a problem of type (7.21). Hence, according to Theorem 7.3 (see (7.22)), we see that $\forall \varepsilon > 0 \; \exists \; \tau_0(\varepsilon)$ which satisfies the inequality

$$\left| z_0(\tau_0) - \varphi(y^0, 0) \right| < \frac{\varepsilon}{3}. \tag{7.29}$$

Let us compare problems (7.26) and (7.27). It follows from 1° that the assumptions of Theorem 7.2 on regular perturbations hold for $\tau \leq \bar{\tau}$, where $\bar{\tau}$ is as large as we wish and fixed. The solution of the unperturbed problem (7.27) is defined for all τ and therefore, in particular, on $[0, \tau_0]$. Hence, according to Theorem 7.2 (together with Remark 2), we see that $\forall \varepsilon > 0$ for sufficiently small μ on $0 \leq \tau \leq \tau_0$ (i.e. on $0 \leq t \leq \tau_0 \mu$) there exists a solution of problem (7.26) (or, which is the same thing, a solution $z(t, \mu), y(t, \mu)$ of problem (7.14), (7.15) and we have the inequalities

$$\left| z(t, \mu) - z_0(\tau) \right| < \frac{\varepsilon}{3}, \quad \left| y(t, \mu) - y^0 \right| < \frac{\varepsilon}{3}. \tag{7.30}$$

Taking into consideration the continuity of $\varphi(y, t)$, it is possible, by choosing the difference $\left| y(t, \mu) - y_0 \right|$ sufficiently small, also to guarantee the inequality

$$\left| \varphi(y(t, \mu), t) - \varphi(y^0, 0) \right| < \frac{\varepsilon}{3} \quad \text{for} \quad 0 \leq t \leq \tau_0 \mu \tag{7.31}$$

It follows from (7.29)–(7.31) that whenever $t = \tau_0 \mu = t_0(\mu)$, we have

$$\left| z(t, \mu) - \varphi(y(t, \mu), t) \right| < \varepsilon. \tag{7.32}$$

2) Denote $\varDelta(t, \mu) = z(t, \mu) - \varphi(y(t, \mu), t)$.

Relation (7.32) shows that $\varDelta(t, \mu)$ is as small as we wish for $t = t_0(\mu)$. Inequality (7.32) will remain valid in some neighbourhood to the right of the point t_0. The size of this neighbourhood is not known in advance. It may turn out that inequality (7.32) is valid for all $t \geq t_0$ up to $t = T$, but it may also turn out that there is some value $t_1 \leq T$ for which (7.32) becomes an equality. Let us check that, for all $t \in [t_0, T]$ in the first case, and for all $t \in [t_0, t_1]$ in the second, the expression $y(t, \mu)$ is as near as we wish to $\bar{y}(t)$. To do this, rewrite the first equation (7.14) in the form

$$\frac{dy}{dt} = f(\varphi(y, t) + \varDelta(t, \mu), y, t),$$

$$y|_{t = t_0(\mu)} = y^0 + \delta(\mu), \tag{7.33}$$

and compare this problem with problem (7.17). According to the results of the previous subsection, the expressions $t_0(\mu)$, $|\delta(\mu)|$ are as small as we wish for a sufficiently small μ and the value $|\varDelta(t,\mu)|$ is as small as we wish for a sufficiently small μ and $t \in [t_0, T]$ or $t \in [t_0, t_1]$. The problem (7.33) is a regular perturbed problem with respect to problem (7.17) (see Theorem 7.2, together with Remark 3). Hence for $t \in [t_0, T]$ or $t \in [t_0, t_1]$ the solution $y(t_1, \mu)$ exists, belongs to D and, moreover, $\forall \varepsilon_1 > 0$ we have the inequality $|y(t, \mu) - \bar{y}(t)| < \varepsilon_1$ when $t \in [t_0, T]$ or $t \in [t_0, t_1]$ as long as μ is sufficiently small.

3) Let us now check that inequality (7.32) holds for all $t \in [t_0, T]$, i.e. of the two possibilities indicated in 2) only one actually takes place, while the assumption on the existence of a point $t_1 \leq T$ for which $|\varDelta(t, \mu)| = \varepsilon$ leads to a contradiction.

Suppose that we have the equality $|\varDelta(t, \mu)| = \varepsilon$ for $t_1 \leq T$. Introduce the function

$$V(z, y, t) = [z - \varphi(y, t)]^2.$$

In view of the assumption concerning the point t_1, we have $\dfrac{dV}{dt}\Big|_{t=t_1} \geq 0$. But

$$\frac{dV}{dt} = 2[z - \varphi(y, t)] \left\{ \frac{F(z, y, t)}{\mu} - \frac{\partial \varphi}{\partial y} f(z, y, t) - \frac{\partial \varphi}{\partial t} \right\}.$$

Since $F(z, y, t)|_{t=t_1} \neq 0$, the sign of $\{\cdot\}$ for sufficiently small μ is determined by the sign of $F(z, y, t)$ and therefore, since $\varDelta$ is small, by the sign of

$$\frac{\partial F}{\partial z}(\varphi(y, t), y, t) [z - \varphi(y, t)]^2.$$

Thus the sign of the entire expression $\dfrac{dV}{dt}$ is determined by the sign of $\dfrac{\partial F}{\partial z}(\varphi(y, t), y, t)$. Since, according to the results of the previous subsection, $y(t, \mu)$ belongs to D, it follows from $3°$ and (7.18) that

$$\frac{\partial F}{\partial z}(\varphi(y(t_1, \mu), t_1), y(t_1, \mu), t_1) < 0$$

and therefore $\dfrac{dV}{dt}\Big|_{t=t_1} < 0$.

This contradiction leads to the conclusion that $|\varDelta(t, \mu)| < \varepsilon$ for $t_0 \leq t \leq T$ and therefore $|y(t, \mu) - \bar{y}(t)| < \varepsilon_1$ for $t_0 \leq t \leq T$. Taking into consideration both of these inequalities, and also the fact that $|y(t, \mu) - y^0| < \varepsilon/3$ for $0 \leq t \leq t_0$ (see (7.30)) and $|\bar{y}(t) - y^0| < \varepsilon_1$ for $0 \leq t \leq t_0$, we obtain the statement of Theorem 7.4.

Remarks. 1. It is clear from the proof that the passage to the limit in (7.24) is uniform on $[0, T]$, while the passage to the limit in (7.25), valid on $(0, T)$, is not uniform. In a neighbourhood of $t = 0$ there is a domain in which the z-component of the solution of problem (7.14), (7.15) greatly differs from the z-component of the degenerate problem's solution, i.e. from $\varphi(\bar{y}(t), t)$. This domain, which is quite visible on Fig. 21 for the case of system (7.21), is referred to as the *boundary layer*.

Also note (this is also clear from the proof itself) that on the segment $[t_0(\mu), T]$ (and hence on any segment $[t_0, T]$, where t_0 is as small as we wish, but is fixed as $\mu \to 0$) both passages to the limit (7.24) and (7.25) are uniform.

2. It is clear from the proof of the theorem that the assumption $\dfrac{\partial F}{\partial z} < 0$ is actually used only for the degenerate solution, i.e. it is sufficient to have the inequality $\dfrac{\partial F}{\partial z}(\varphi(\bar{y}(t), t), \bar{y}(t), t) < 0$.

3. Theorem 7.4 can be generalized to the vector case. Certain supplementary difficulties arise with the definition of the domain of attraction. Concerning the stability condition, it can be stated as a natural requirement that the characteristic roots $\lambda(t)$ of the matrix $\dfrac{\partial F}{\partial z}(\varphi(\bar{y}, t), \bar{y}, t)$ satisfy the inequality $\operatorname{Re} \lambda < 0$. There is also a more general statement of the stability condition, given by A. N. Tikhonov (see [23]).

2. Asymptotic expansion of the solution of problem (7.14), (7.15). Using Theorem 7.4, it is possible to write the following asymptotic representation for the solution of problem (7.14), (7.15):

$$y(t, \mu) = \bar{y}(t) + \varepsilon_1(t, \mu), \quad z(t, \mu) = \bar{z}(t) + \varepsilon_2(t, \mu) \quad (\bar{z}(t) \equiv \varphi(\bar{y}(t), t)).$$

Here the remainder $\varepsilon_2(t, \mu)$ and the expression for $z(t, \mu)$ are not uniformly small magnitudes. It is natural to assume that if we add the difference $z_0(\tau) - \varphi(y^0, 0)$ to $\bar{z}(t)$, then we obtain the formula

$$z(t, \mu) = \bar{z}(t) + z_0(\tau) - \varphi(y^0, 0) + \varepsilon_3(t, \mu)$$

where $\varepsilon_3(t, \mu) \to 0$ uniformly on $[0, T]$.

Indeed, we have $\varepsilon_3(t, \mu) = z(t, \mu) - \bar{z}(t) - z_0(\tau) + \varphi(y^0, 0)$. Let us split up the closed interval $[0, T]$ into two parts $[0, t_0(\mu)]$ and $[t_0(\mu), T]$, where $t_0(\mu) = \tau_0 \mu$ is the value which appeared in the proof of theorem 7.4, τ_0 is sufficiently large and fixed when $\mu \to 0$. On $[0, t_0(\mu)]$ represent ε_3 in the form

$$\varepsilon_3(t, \mu) = [z(t, \mu) - z_0(\tau)] - [\bar{z}(t) - \varphi(y^0, 0)].$$

Here $|z(t,\mu)-z_0(\tau)|<\varepsilon/2$ as $\mu\to0$ (this is the same as (7.30)), while $|\bar{z}(t)-\varphi(y^0,0)|=|\bar{z}(t)-(0)|<\varepsilon/2$, since $t_0(\mu)=0$ as $\mu\to0$. Therefore on $[0,t_0(\mu)]$ we have $|\varepsilon_3|<\varepsilon$ uniformly with respect to t as $\mu\to0$. On $[t_0(\mu),T]$ represent ε_3 in the form

$$\varepsilon_3(t,\mu)=[z(t,\mu)-\bar{z}(t)]-[z_0(\tau)-\varphi(y^0,0)].$$

Here $|z(t,\mu)-\bar{z}(t)|<\varepsilon/2$ (see remark 1 to theorem 7.4 concerning the uniformity of the limit in (7.24) on $[t_0(\mu),T]$). Exactly in the same way, $|z_0(\tau)-\varphi(y^0,0)|<\varepsilon/2$, since this is the same inequality as in the particular case (7.23), i.e. on $[t_0(\mu),T]$ we also have $|\varepsilon_3|<\varepsilon$ uniformly with respect to t as $\mu\to0$. Thus $\varepsilon_3(t,\mu)\to0$ uniformly as $\mu\to0$ on $[0,T]$.

Note that the difference $z_0(\tau)-\varphi(y^0,0)$ plays the role of the correction to the "loss" of the initial condition $z(0,\mu)=z^0$, which cannot be satisfied by $\bar{z}(t)$. The expression $z(t)+z_0(\tau)-\varphi(y^0,0)$ will already satisfy the initial condition for $t=0$.

Further (Theorem 7.5) we shall prove that

$$\varepsilon_1(t,\mu)=O(\mu),\quad \varepsilon_3(t,\mu)=O(\mu).$$

Moreover, for sufficiently smooth right-hand side in (7.14) we can construct an asymptotic representation for the solution of problem (7.14), (7.15) with residual term $O(\mu^{M+1})$ but, unlike the regular case (see Theorem 7.1), this representation, besides terms with powers of μ (or *regular terms*), will contain certain functions (the so-called *boundary terms*) in which μ appears in non-polynomial fashion; the boundary terms have a large value in the neighbourhood of $t=0$ and rapidly decrease with the increase of t. The difference $z_0(\tau)-\varphi(y^0,0)$ introduced above is a boundary term in the asymptotic formula with remainder $O(\mu)$.

After these preliminary remarks, let us pass to the direct description of the general algorithm for constructing the asymptotics of solutions of the singularly perturbed problem (7.14), (7.15).

Represent z and y in the form of the sum of two formal series (here and later by x we shall mean z and y taken together, i.e. if we write out some relation for x, then this means that we have exactly the same relation for both z and y)

$$x=\bar{x}(t,\mu)+\Pi x(\tau,\mu),\tag{7.34}$$

where

$$\bar{x}(t,\mu)=\bar{x}_0(t)+\mu\bar{x}_1(t)+\dots\tag{7.35}$$

is the so-called *regular series* (compare with (7.6)), i.e. a series in powers of μ with coefficients depending on t, while

$$\Pi x(\tau,\mu)=\Pi_0 x(\tau)+\mu\Pi_1 x(\tau)+\dots\tag{7.36}$$

is the so-called *boundary series*, which is also a series in powers of μ, but whose coefficients depend on τ. Terms of this series are called *boundary terms*.

Substituting (7.34) into (7.14) and multiplying the second equation by μ for symmetry's sake, we get

$$\mu \frac{d\bar{z}}{dt} + \frac{d}{d\tau} \, \Pi z = F(\bar{z} + \Pi z, \; \bar{y} + \Pi y, t),$$

$$\mu \frac{d\bar{y}}{dt} + \frac{d}{d\tau} \, \Pi y = \mu f(\bar{z} + \Pi z, \; \bar{y} + \Pi y, t). \tag{7.37}$$

Introduce the notations $\bar{F}$ and ΠF by putting

$$\bar{F} = F(\bar{z}(t, \mu), \; \bar{y}(t, \mu), t),$$

$$\Pi F = F(\bar{z}(\tau\mu, \mu) + \Pi z(\tau, \mu), \; \bar{y}(\tau\mu, \mu) + \Pi y(\tau, \mu), \tau\mu)$$

$$- F(\bar{z}(\tau\mu, \mu), \; \bar{y}(\tau\mu, \mu), \tau\mu)$$

(similar notations $\bar{f}$ and Πf are introduced for f). Then (7.37) can be written in the form

$$\mu \frac{d\bar{z}}{dt} + \frac{d}{d\tau} \, \Pi z = \bar{F} + \Pi F, \quad \mu \frac{d\bar{y}}{dt} + \frac{d}{d\tau} \, \Pi y = \mu(\bar{f} + \Pi f). \tag{7.38}$$

Now expand $\bar{F}$ and ΠF formally in powers of μ (the coefficients in these decompositions will depend on t and τ respectively):

$$\bar{F} = \bar{F}_0 + \mu \bar{F}_1 + \dots, \qquad \Pi F = \Pi_0 F + \mu \Pi_1 F + \dots$$

and set the coefficients of identical powers of μ equal to each other in (7.38), those depending on t and those depending on τ separately:

$$\frac{d}{dt} \, \bar{z}_{k-1} = \bar{F}_k, \quad \frac{d}{dt} \, \bar{y}_k = \bar{f}_k; \tag{7.39}$$

$$\frac{d}{d\tau} \, \Pi_k z = \Pi_k F, \quad \frac{d}{d\tau} \, \Pi_k y = \Pi_{k-1} f. \tag{7.40}$$

Thus we have obtained equations determining the terms of the expansions (7.35) and (7.36).

Let us write out these equations in more detail for $k = 0$. We have

$$0 = \bar{F}_0 \equiv F(\bar{z}_0, \bar{y}_0, t), \quad \frac{d\bar{y}_0}{dt} = f(\bar{z}_0, \bar{y}_0, t). \tag{7.41}$$

This system coincides, as could be expected, with the degenerate system (7.16).

We also have (taking into consideration the first of the equations (7.41))

$$\frac{d}{d\tau}\,\Pi_0 z = \Pi_0 F \equiv F(\bar{z}_0(0) + \Pi_0 z, \bar{y}_0(0) + \Pi_0 y, 0) - F(\bar{z}_0(0), \bar{y}_0(0), 0)$$

$$\equiv F(\bar{z}_0(0) + \Pi_0 z, \bar{y}_0(0) + \Pi_0 y, 0), \qquad \frac{d}{d\tau}\,\Pi_0 y = 0. \qquad (7.42)$$

Beginning with $k = 1$, equations (7.39) and (7.40) will be linear with respect to $\bar{z}_k, \bar{y}_k$ and $\Pi_k z, \Pi_k y$. Note that system (7.39) does not contain any derivative of $\bar{z}_k$ but only derivatives of $\bar{y}_k$, while the system (7.40) has the following special property: the second equation in it separates from the others, since its right-hand side only contains terms of previous numbers.

In order to determine the terms of the expansions (7.35), (7.36) from the equations (7.39), (7.40) obtained above, it is necessary to fix the initial conditions. To do this, let us substitute (7.34) into (7.15):

$$\bar{x}_0(0) + \mu \bar{x}_1(0) + \ldots + \Pi_0 x(0) + \mu \Pi_1 x(0) + \ldots = x^0 \qquad (7.43)$$

and set the coefficients of identical powers of μ in both parts of these equations equal to each other. When we set the zero degree terms equal to each other, we get

$$\bar{z}_0(0) + \Pi_0 z(0) = z^0, \qquad \bar{y}_0(0) + \Pi_0 y(0) = y^0. \qquad (7.44)$$

Let us consider the second one of these equations. Without supplementary considerations it is impossible to determine $\bar{y}_0(0)$ and $\Pi_0 y(0)$ from it. However, the boundary terms, which will be called upon to play the role of corrections in the neighbourhood $t = 0$, and which for $t > 0$ must tend to zero together with μ must satisfy the supplementary condition $\Pi_i x \to 0$ as $\tau \to \infty$. Hence we come to the conclusion that $\Pi_0 y(0) = 0$ since otherwise (see (7.42)) $\Pi_0 y(\tau) \equiv \Pi_0 y(0) = \text{const} \nrightarrow 0$. But then it follows from (7.44) that

$$\bar{y}_0(0) = y^0. \qquad (7.45)$$

Solving system (7.41) under this condition, we see that $\bar{z}_0(t), \bar{y}_0(t)$ coincides with the solution $\bar{z}(t), \bar{y}(t)$ which was already obtained in Theorem 7.4. From (7.41) we get $z_0(0) = \varphi(\bar{y}_0(0), 0) = \varphi(y^0, 0)$ and then the first of the relations (7.44) gives us an initial condition for $\Pi_0 z$. Thus the initial conditions for the system (7.42) are of the form

$$\Pi_0 y(0) = 0, \ \Pi_0 z(0) = z^0 = \bar{z}_0(0) = z^0 - \varphi(y^0, 0). \qquad (7.46)$$

Hence $\Pi_0 y(\tau) \equiv 0$ and $\Pi_0 z(\tau)$ is the solution of the following initial value problem

$$\frac{d}{d\tau}\,\Pi_0 z = F(\varphi(y^0, 0) + \Pi_0 z, y^0, 0)$$

$$\Pi_0 z(0) = z^0 - \varphi(y^0, 0). \qquad (7.47)$$

Comparing problem (7.47) with problem (7.28) it is easy to establish that $\Pi_0 z = \bar{z}_0(\tau) - \varphi(y^0, 0)$. This difference was already considered previously before we began to describe the general algorithms. The inequality (7.29) obtained above means that $\Pi_0 z(\tau) \to 0$ as $\tau \to \infty$.

Thus the zero degree terms of the expansions (7.35), (7.36) are entirely determined.

Setting the coefficients of the first degree terms in μ in (7.43) equal to each other, we obtain

$$\bar{z}_1(0) + \Pi_1 z(0) = 0, \qquad \bar{y}_1(0) + \Pi_1 y(0) = 0. \tag{7.48}$$

We must also make use of the condition $\Pi_1 y \to 0$ as $\tau \to \infty$. From the second equation (7.40) (for $k = 1$) we get

$$\Pi_1 y(\tau) - \Pi_1 y(0) = \int\limits_0^\tau \Pi_0 f d\tau,$$

hence by the condition $\Pi_1 y \to 0$ as $\tau \to \infty$, it follows that

$$\Pi_1 y(0) = - \int\limits_0^\infty \Pi_0 f d\tau. \tag{7.49}$$

The convergence of the integral which appears here will be proved later (see (7.54)) and in general all our calculations at this stage are purely formal. If we take into consideration (7.49) for $\Pi_1 y$, we get

$$\Pi_1 y(\tau) = - \int\limits_\tau^\infty \Pi_0 f d\tau. \tag{7.50}$$

The second relation (7.48) now implies

$$\bar{y}_1(0) = \int\limits_0^\infty \Pi_0 f d\tau. \tag{7.51}$$

This will be the initial condition for the system (7.39) when $k = 1$, and thus we can determine $\bar{y}_1(t), \bar{z}_1(t)$. After this, the second relation (7.48) yields

$$\Pi_1 z(0) = -\bar{z}_1(0). \tag{7.52}$$

This condition allows us to find $\Pi_1 z$ from the first equation (7.40) when $k = 1$, since $\Pi_1 y$ has already been determined.

In a similar way, we determine $\Pi_k y, \bar{y}_k, \bar{z}_k, \Pi_k z$ $(k = 2, 3, \dots)$ from the system (7.39), (7.40) by using the supplementary conditions

$$\Pi_k x \to 0 \text{ as } k \to \infty,$$

$$\bar{y}_k(0) = \int\limits_0^\infty \Pi_{k-1} f d\tau, \qquad \Pi_k z(0) = -\bar{z}_k(0). \tag{7.53}$$

This concludes our description of the expansion (7.34).

In the theory of singular perturbations, the following estimate for $\Pi_k x$ is proved:

$$|\Pi_k x| < C e^{-\kappa t/\mu} \quad (k = 0, 1, \ldots), \tag{7.54}$$

where $C > 0$, $\kappa > 0$ are certain constants. This estimate means that $\Pi_k x$ tends to zero exponentially as $\tau \to \infty$, and this same inequality guarantees the convergence of the integral in (7.53).

The main statement concerning the constructions just carried out is that the series (7.34) is the asymptotic series for the solution $x(t, \mu)$ of problem (7.14), (7.15); it is proved in the theory of singular perturbations that the difference between $x(t, \mu)$ and the n-th partial sum of the series (7.34) is of order $O(\mu^{n+1})$. This is the generalization of theorem 7.1 to singularly perturbed systems. This is described in more detail in the book [23].

Let us give a proof of the validity of the asymptotic expansion for the solution of problem (7.14), (7.15) with remainder $O(\mu)$, namely let us prove that

$$z(t, \mu) = \bar{z}_0(t) + \Pi_0 z + O(\mu), \quad y(t, \mu) = \bar{y}_0(t) + O(\mu).$$

The proof of the validity of this expansion with remainder $O(\mu^{n+1})$ is more difficult only in its purely technical aspects.

Let us give an exact statement of the corresponding theorem.

Theorem 7.5. *Suppose we have the following assumptions:*

1°. $F(z, y, t)$ and $f(z, y, t)$ are continuous with respect to the set of their variables in some domain H.

2°. On the closed interval $[0, T]$ the solution $\bar{y}_0(t)$, $\bar{z}_0(t)$ of the problem (7.41), (7.45) is defined and this solution belongs to H.

3°. $F_z(\bar{z}_0(t), \bar{y}_0(t), t)$ exists, is continuous and is negative when $t \in [0, T]$.

4°. z^0 belongs to the domain of attraction of $\varphi(y^0, 0)$.

5°. When $0 \le t \le T, |\Delta| < \varepsilon, |\delta| < \varepsilon$ (ε is a sufficiently small but fixed number which does not depend on μ)

$$F(\bar{z}_0 + \Pi z + \Delta, \bar{y}_0 + \delta, t), f(\bar{z}_0 + \Pi z + \Delta, \bar{y}_0 + \delta, t)$$

are continuous together with their derivatives with respect to Δ and δ up to the second order inclusive.

Then, on the interval $0 \le t \le T$, we have the following estimates, which are uniform with respect to $t \in [0, T]$

$$z(t, \mu) - \bar{z}_0(t) - \Pi_0 z \left(\frac{t}{\mu}\right) = O(\mu) \tag{7.55}$$

$$y(t, \mu) - \bar{y}_0(t) = O(\mu). \tag{7.56}$$

The proof of the theorem is based on a series of lemmas.

Lemma 7.1. *We have the inequality*

$$|\Pi_0 z| < C e^{-\kappa t \mu}, \tag{7.57}$$

where $C > 0$, $\kappa > 0$ are certain constants.

We have already seen above that $\Pi_0 z \to 0$ as $\tau \to \infty$. Now it is only necessary to specify how this expression tends to zero. Note that the relation (7.57) is (7.54) for $x = z$, $k = 0$. It follows from (7.47) (recall that $\varphi(y^0, 0) = \bar{z}_0(0)$, $y^0 = \bar{y}_0(0)$) that

$$\frac{d}{d\tau} \Pi_0 z = F_z(\bar{z}_0(0) + \theta\Pi_0 z, \bar{y}_0(0), 0)\Pi_0 z,$$

hence

$$\Pi_0 z = [z^0 - \varphi(y^0, 0)] e^{\int_0^\tau F_z(\bar{z}_0(0) + \theta\Pi_0 z, \bar{y}_0(0), 0)d\tau}$$

Choose τ_0 so that for $\tau \geq \tau_0$ the value of $\Pi_0 z$ is small enough and fix this number τ_0. The value of $\Pi_0 z$ must be so small that, as a consequence of $3°$, we have the inequality

$$F_z(\bar{z}_0(0) + \theta\Pi_0 z, \bar{y}_0(0), 0) < -\kappa < 0.$$

Then we have

$$|\Pi_0 z| = |z^0 - \varphi(y^0, 0)| e^{\int_0^{\tau_0} F_z d\tau} e^{\int_{\tau_0}^\tau F_z d\tau} < |z^0 - \varphi(y^0, 0)| e^{\int_0^{\tau_0} F_z d\tau} e^{-\kappa(\tau - \tau_0)} < C^{-\kappa\tau},$$

since $|z^0 - \varphi(y^0, 0)| e^{\int_0^{\tau_0} F_z d\tau} e^{\kappa\tau_0} < C$.

Substituting into (7.14), (7.15) the principal terms of the asymptotic expansion, i.e. the expressions $\bar{z}_0 + \Pi_0 z$, $\bar{y}_0$ instead of the exact solution, we get

$$\mu \frac{d\bar{z}_0}{dt} + \frac{d}{d\tau} \Pi_0 z = F(\bar{z}_0(t) + \Pi_0 z, \bar{y}_0(t), t) - R_1$$

$$\frac{d\bar{y}_0}{dt} = f(\bar{z}_0(t) + \Pi_0 z, \bar{y}_0(t), t) - R_2, \tag{7.58}$$

where $-R_1$, $-R_2$ are the corresponding discrepancies.

Lemma 7.2. *The following estimates hold*

$$R_1(t, \mu) = O(\mu), \qquad R_2(t, \mu) = O(e^{-\kappa t/\mu}).$$

The estimate for R_2 is obtained immediately from Lemma 7.1, since, if we take into consideration (7.41), we get $R_2 = f_z(\bar{z}_0 + \theta\Pi_0 z, \bar{y}_0, t)\Pi_0 z$.

In order to obtain the estimate for R_1, it is sufficient check that

$$R_{(1)1} = F(\bar{z}_0 + \Pi_0 z, \bar{y}_0, t) - \frac{d}{dt} \Pi_0 z = O(\mu).$$

Using (7.47), let us represent $R_{(1)1}$ in the form

$$R_{(1)1} = F(\bar{z}_0(t) + \Pi_0 z, \bar{y}_0(t), t) - F(\bar{z}_0(0) + \Pi_0 z, \bar{y}_0(0), 0)$$

$$= \Phi(\Pi_0 z, t) - \Phi(\Pi_0 z, 0) = t \int_0^1 \Phi_t(\Pi_0 z, \theta t) d\theta.$$

But

$$\Phi_t(\Pi_0 z, \theta t) = F_z(\bar{z}_0(\theta t) + \Pi_0 z, \bar{y}_0(\theta t), \theta t) \bar{z}_0'(\theta t) + F_y(\cdot) \bar{y}_0'(\theta t) + F_t(\cdot)$$

$$= F_{zz}(\bar{z}_0(\theta t) + \theta_1 \Pi_0 z, \bar{y}_0(\theta t), \theta t) \bar{z}_0'(\theta t) \Pi_0 z.$$

Here we have taken into consideration the fact that $F(\bar{z}_0, \bar{y}_0, t) = 0$ and therefore

$$F_z(\bar{z}_0, \bar{y}_0, t) \bar{z}_0' + F_y(\cdot) \bar{y}_0' + F_t(\cdot) = 0.$$

Hence

$$|R_{(1)1}| < t|F_{zz}||z_0'||\Pi_0 z| < Cte^{-\kappa t/\mu} < C\mu,$$

since $\sup_t (te^{-\kappa t/\mu}) = \mu e^{-1}/\kappa$.

Further constant Ce^{-1}/κ is again denoted by C. In the sequel, let us agree to denote all the constants which do not depend on μ by the same letter C, except in some special cases, as earlier in §2, Chapter 5.

Now let us begin the proof of Theorem 7.5. Put $\Delta = z - \bar{z}_0 - \Pi_0 z$, $\delta = y - \bar{y}_0$. Subtracting (7.58) from (7.14), we get

$$\mu \frac{d\Delta}{dt} = F(\bar{z}_0 + \Pi_0 z + \Delta, \bar{y}_0 + \delta, t) - F(\bar{z}_0 + \Pi_0 z, \bar{y}_0, t) + R_1$$

$$\frac{d\delta}{dt} = f(\bar{z}_0 + \Pi_0 z + \Delta, \bar{y}_0 + \delta, t) - f(\bar{z}_0 + \Pi_0 z, \bar{y}_0, t) + R_2. \quad (7.59)$$

Obviously,

$$\Delta(0, \mu) = 0, \quad \delta(0, \mu) = 0. \quad (7.60)$$

The system of equations 7.59 can be rewritten in the form

$$\mu \frac{d\Delta}{dt} = a_{11}\Delta + a_{12}\delta + R_1$$

$$\frac{d\delta}{dt} = a_{21}\Delta + a_{22}\delta + R_2 \quad (7.61)$$

where

$$a_{11}(\Delta, \delta, t, \mu) = \int_0^1 F_z(\bar{z}_0 + \Pi_0 z + \theta\Delta, \bar{y}_0 + \delta, t) d\theta$$

$$a_{12}(\Delta, \delta, t, \mu) = \int_0^1 F_y(\bar{z}_0 + \Pi_0 z, \bar{y}_0 + \theta\delta, t) d\theta,$$

a_{21}, a_{22} have the same structure. The fact that the right-hand sides of (7.59) and (7.61) are equal can be checked directly if we note that in the expression for a_{11} under the integral sign we actually have the value $\dfrac{1}{\Delta}\dfrac{dF}{d\theta}(\bar{z}_0+\Pi_0 z+\theta\Delta,$ $\bar{y}_0+\delta,t)$ and then apply similar considerations to the other coefficients a_{ik}.

Using the first equation (7.61), let us express Δ in terms of δ (in the sequel we shall use the notation $\Delta(t)$, $\delta(t)$, omitting to write out the dependence on μ explicitly)

$$\Delta(t)=\int_0^t \frac{e^{\frac{1}{\mu}\int_\xi^t a_{11}dt}}{\mu}\,(a_{12}\delta(\xi)+R_1)d\xi \tag{7.62}$$

and substitute this into the second equation:

$$\frac{d\delta}{dt}=a_{22}\delta+a_{21}\int_0^t \frac{e^{\frac{1}{\mu}\int_\xi^t a_{11}dt}}{\mu}\,(a_{12}\delta(\xi)+R_1)d\xi+R_2.$$

Thus we obtain

$$\delta(t)=\int_0^t e^{\int_\xi^t a_{22}dt}R_2 d\xi+\int_0^t e^{\int_\xi^t a_{22}dt}d\xi\, a_{21}\int_0^\xi \frac{e^{\frac{1}{\mu}\int_\eta^\xi a_{11}dt}}{\mu}\,(a_{12}\delta(\eta)+R_1)d\eta.$$

Changing the order of integration in the second summand, we can write this equation in the following form

$$\delta(t)=\int_0^t \mathscr{K}(t,\eta,\mu)\delta(\eta)d\eta+R, \tag{7.63}$$

where $\mathscr{K}(t,\eta,\mu)=a_{12}q(t,\eta,\mu)$,

$$q(t,\eta,\mu)=\int_\eta^t a_{21}\,e^{\int_\xi^t a_{22}dt}\,\frac{e^{\frac{1}{\mu}\int_\eta^\xi a_{11}dt}}{\mu}\,d\xi \tag{7.64}$$

$$R=\int_0^t e^{\int_\xi^t a_{22}dt}R_2 d\xi+\int_0^t R_1 q\, d\eta. \tag{7.65}$$

It should be noted that the expression $\mathscr{K}(t,\eta,\mu)$ in equations (7.63) actually depends on Δ and δ, since a_{ik} depends on Δ and δ, but expressions Δ and δ are viewed as known functions of t.

Now let us estimate the values of q and R, using the fact (already proved at the beginning of this subsection) that Δ and δ uniformly tend to zero.

First estimate q. Since Δ and δ tend to zero uniformly as $\mu\to0$, we obtain $a_{11}=F_z(\bar{z}_0+\Pi_0 z,\bar{y}_0,t)+\gamma(t,\mu)$ where $\gamma(t,\mu)\to0$ uniformly as $\mu\to0$. Now using considerations similar to those applied for the estimate of $\Pi_0 z$ in lemma 7.1, it is easy to get

$$e^{\frac{1}{\mu}\int_\eta^\xi a_{11}dt}<Ce^{-\frac{\kappa_1}{\mu}(\xi-\eta)}. \tag{7.66}$$

Indeed, according to 3°, we have $F_z(\bar{z}_0(t), \bar{y}_0(t), t) < 0$. Therefore, as in Lemma 7.1, since $\Pi_0 z$ is small when $t \geq \tau_0 \mu$ (where τ_0 is sufficiently large) while $\mu \to 0$)we get $a_{11} < -\kappa_1 < 0$ by using the fact that $\gamma(t, \mu)$ is uniformly small when $\mu \to 0$. The rest is a word for word repetition of the arguments proving Lemma 7.1.

Thus we have (7.66). Furthermore it is obvious that the first exponential factor in the expression under the integral sign is bounded from above in absolute value by a certain constant C (which does not depend on μ) and therefore

$$|q| < \int_\eta^t C \, \frac{e^{-\frac{\kappa_1}{\mu}(\xi - \eta)}}{\mu} \, d\xi < C. \tag{7.67}$$

Now let us estimate R. In view of Lemma 7.2, we have

$$|R| < \int_0^t Ce^{-\frac{\kappa\xi}{\mu}} d\xi + \int_0^t C\mu d\eta < C\mu. \tag{7.68}$$

In other words, $R = O(\mu)$.

Now let us consider equations (7.63). Using the estimates (7.67) and (7.68), we obtain the inequality

$$|\delta| < C \int_0^t |\delta| d\eta + C\mu. \tag{7.69}$$

Introduce $u = \int_0^t |\delta| d\eta$. Then $\frac{du}{dt} = |\delta|$, $\left|\frac{du}{dt}\right| = |\delta|$ and (7.69) gives us

$$\left|\frac{du}{dt}\right| < C|u| + C\mu. \tag{7.70}$$

Using Lemma 2.1 on differential inequalities, we obtain

$$|u| < \mu(e^{Ct} - 1) \leq \mu(e^{CT} - 1) < C\mu. \tag{7.71}$$

Therefore by (7.70) we have $\frac{du}{dt} < C\mu$, i.e.

$$|\delta| < C\mu. \tag{7.72}$$

Now using (7.62), the estimate (7.72), the fact that $|R_1| < C\mu$ and the estimate (7.66), we obtain

$$|\Delta| < \int_0^t \frac{e^{-\frac{\kappa_1}{\mu}(t - \xi)}}{\mu} \, C\mu d\xi < C\mu. \tag{7.73}$$

The estimates (7.72) and (7.73) constitute the statement of theorem 7.5, which is therefore proved.

3. Constructing the asymptotics of the fundamental system of solutions for linear second order equations with small parameter multiplying the highest order derivatives. In a series of problems of oscillation theory, quantum mechanics and others, one meets with a singularly perturbed equation of the form

$$\mu^2 y'' + Q^2(x) y = 0 \tag{7.74}$$

which does not satisfy the requirements of the previous subsection. In particular, the equation of a pendulum (7.11) without friction, i.e. in the case $\alpha = 0$, relates to the same type as (7.74). In order to reduce (7.11) to a system of type (7.14) in which we would have the stability condition (7.18) – the basis of the entire theory in subsection 1 – it is necessary that α differ from zero. In the case $\alpha = 0$, however, the theory of subsection 1 cannot be applied. As is well known the solution of equation (7.11) in this case will be oscillatory in character (in view of the fact that μ is small, the oscillations will have very high frequency), i.e. the phenomenon described is qualitatively different from the one considered in subsection 1.

Using the linearity of equation (7.74), we shall not relate the construction of the asymptotics with the choice of certain supplementary conditions, as was done in the previous subsection (where we considered an initial value problem), but will attempt to construct the asymptotics of the fundamental system of solutions; this will enable us to obtain the asymptotics of solutions satisfying diverse supplementary conditions.

We shall assume that on the closed interval $a \leq x \leq b$ we have $Q(x) \neq 0$ and this function possesses continuous derivatives to the third order (to be definite, assume that $Q(x) > 0$).

In equation (7.74) let us pass to a new unknown function u by putting

$$y = uv, \quad \text{where} \quad v = \exp\left[\frac{i}{\mu} \int_a^x Q(x) dx\right]. \tag{7.75}$$

Note that in the case $Q = \text{const}$, v becomes $\exp\left[\frac{i}{\mu} Q(x - a)\right]$ and the expression $y = uv$ (where $u = \text{const}$) is simply an exact solution.

Carrying out the substitution indicated above in (7.74), we obtain the equation $\mu u'' + 2iu'Q + iuQ' = 0$ which can be written in the form of the system

$$\mu z' = -2izQ - iuQ', \quad u' = z. \tag{7.76}$$

Taking $\mu = 0$, we obtain $2zQ = -uQ'$, $u' = z$, hence

$$u' = -\frac{1}{2}\frac{Q'}{Q} u.$$

Consider the particular solution of this equation of the form

$$u(x) = \frac{1}{\sqrt{Q(x)}}. \tag{7.77}$$

It turns out that if we substitute $\bar{u}(x)$ into (7.75) in place of u, then we obtain an approximate (in the asymptotic sense) representation for a certain solution of equation (7.74) which we shall denote by $y_1(x)$. Denote by $y_2(x)$ the solution complex conjugate to $y_1(x)$: $y_2 = y_1^*$.

Theorem 7.6. *Suppose the function $Q(x) > 0$ is positive and possesses a continuous third derivative on $[a, b]$. Then there exists a fundamental system of solutions of equation (7.74) on $[a, b]$ of the form*

$$y_1 = \left[\frac{1}{\sqrt{Q(x)}} + \varepsilon_1(x, \mu)\right] \exp\left[\frac{i}{\mu} \int_a^x Q(x)dx\right],$$

$$y_2 = \left[\frac{1}{\sqrt{Q(x)}} + \varepsilon_2(x, \mu)\right] \exp\left[-\frac{i}{\mu} \int_a^x Q(x)dx\right]. \tag{7.78}$$

and $\varepsilon_1(x, \mu) = O(\mu)$, $\varepsilon_2(x, \mu) = \varepsilon_1^* = O(\mu)$.

To prove Theorem 7.6, let us put $u - \bar{u} = \delta$, $z - \bar{z} = \Delta$ where $\bar{u}$ is defined by formula (7.77) and $\bar{z} = \bar{u}'$. For Δ and δ, we obtain the system

$$\mu\Delta' = -2iQ\Delta - iQ'\delta - \mu\bar{z}', \qquad \delta' = \Delta \tag{7.79}$$

and can determine these functions by setting the initial conditions

$$\delta(a) = 0, \qquad \Delta(a) = 0 \tag{7.80}$$

(thus we actually give initial conditions for u and z).

The proof shall be split up into two parts.

1) Let us first prove that

$$\delta = O(\mu), \qquad \Delta = O(\mu). \tag{7.81}$$

(7.81) means that there exist solutions of the system (7.76) of the form $u = \bar{u} + O(\mu)$, $z = \bar{z} + O(\mu)$ and therefore there is a solution y_1 of equation (7.74), represented by the first formula (7.78). As for the second solution y_2, since $y_2 = y_1^*$, the representation for y_2 does not necessitate a special proof. Here $\varepsilon_2(x, \mu) = \varepsilon_1^*(x, \mu)$.

To prove (7.81), let us pass from (7.79), (7.80) to the equivalent system of integral equations

$$\Delta(x) = \int_a^x \frac{-iQ'(\tau)\delta(\tau) - \mu\bar{z}'(\tau)}{\mu} \, e^{-\frac{2i}{\mu}\int_\tau^x Qdx} \, d\tau,$$

$$\delta(x) = \int_a^x \Delta(\xi)d\xi, \tag{7.82}$$

and then to an equation containing $\Delta(x)$ only

$$\Delta(x) = \int_a^x \frac{-iQ'(\tau)}{\mu} \left(\int_a^\tau \Delta(\xi)d\xi \right) e^{-\frac{2i}{\mu}\int_\tau^x Qdx} \, d\tau - \int_a^x \bar{z}'(\tau)e^{-\frac{2i}{\mu}\int_\tau^x Qdx} \, d\tau = (1) + (2). \tag{7.83}$$

Integrating the second summand of this expression by parts, we obtain

$$(2) = -\mu e^{-\frac{2i}{\mu}\int_\tau^x Qdx} \left. \frac{\bar{z}'(\tau)}{2iQ} \right|_a^x + \mu \int_a^x e^{-\frac{2i}{\mu}\int_\tau^x Qdx} \left(\frac{\bar{z}'}{2iQ} \right)' d\tau = \alpha(x,\mu)\mu.$$

Now transforming (1) by a similar integration by parts, we get

$$(1) = \left(-\frac{Q'(\tau)}{2Q(\tau)} \int_a^\tau \Delta(\xi)d\xi \right) e^{-\frac{2i}{\mu}\int_\tau^x Qdx} \bigg|_a^x$$

$$+ \int_a^x \left(\frac{Q'(\tau)}{2Q(\tau)} \int_a^\tau \Delta(\xi)d\xi \right)' e^{-\frac{2i}{\mu}\int_\tau^x Qdx} \, d\tau = -\frac{Q'(x)}{2Q(x)} \int_a^x \Delta(\tau)d\tau$$

$$+ \int_a^x \left[\left(\frac{Q'(\tau)}{2Q(\tau)} \right)' \int_a^\tau \Delta(\xi)d\xi + \frac{Q'(\tau)}{2Q(\tau)} \Delta(\tau) \right] e^{-\frac{2i}{\mu}\int_\tau^x Qdx} \, d\tau.$$

Thus, equation (7.83) acquires the form

$$\Delta(x) = \int_a^x \beta(x,\tau,\mu)d\tau \int_a^\tau \Delta(\xi)d\xi + \int_a^x \gamma(x,\tau,\mu)\Delta(\tau)d\tau + \alpha(x,\mu)\mu,$$

and we have $|\alpha(x,\mu)| < \bar{\alpha}$, $|\beta(x,\tau,\mu)| < \bar{\beta}$, $|\gamma(x,\tau,\mu)| < \bar{\gamma}$, where $\bar{\alpha}, \bar{\beta}, \bar{\gamma}$ are certain constants not depending on μ. Changing the order of integration in the first summand, we obtain

$$\int_a^x \Delta(\xi)d\xi \int_\xi^x \beta(x,\tau,\mu)d\tau = \int_a^x \tilde{\beta}(x,\xi,\mu)\Delta(\xi)d\xi,$$

where $\tilde{\beta}(x,\xi,\mu) < \bar{\delta}$, $\bar{\delta}$ is a constant not depending on μ. Thus equation (7.83) has acquired its final form

$$\Delta(x) = \int_a^x K(x, \tau, \mu)\Delta(\tau)d\tau + \alpha(x, \mu)\mu, \qquad (7.84)$$

where $K(x, \tau, \mu) = \bar{\beta}(x, \tau, \mu) + \gamma(x, \tau, \mu)$, and therefore $|K(x, \tau, \mu)| < \bar{K}$ (the constant $\bar{K}$ does not depend on μ).

Replacing equation (7.84) by an inequality, we get

$$|\Delta(x)| < \int_a^x \bar{K}|\Delta(\tau)|d\tau + \bar{\alpha}\mu. \qquad (7.85)$$

Introduce the new function $w(x) = \int_a^x |\Delta(\tau)|d\tau$ (similar constructions were carried out in our proof of Theorem 7.5). Then $\dfrac{dw}{dx} - |\Delta|$, $\left|\dfrac{dw}{dx}\right| - |\Delta|$ and (7.85) implies

$$\left|\frac{dw}{dx}\right| < \bar{K}|w| + \bar{\alpha}\mu.$$

Using Lemma 2.1 on differential inequalities, we obtain

$$|w| < \frac{\bar{\alpha}\mu}{\bar{K}} (e^{\bar{K}(x-a)} - 1) \le \frac{\bar{\alpha}\mu}{\bar{K}} (e^{\bar{K}(b-a)} - 1) < C\mu.$$

Therefore, $|\Delta(x)| = \left|\dfrac{dw}{dx}\right| < \bar{K}C\mu + \bar{\alpha}\mu < \bar{c}\mu$, which was required. A similar estimate for $\delta(x)$ is obtained from the second equation (7.82). Thus the estimates (7.81) are established.

2) Now let us verify that the solutions y_1 and $y_2 = y_1^*$ are linearly independent, thus proving that the Wronskian $D(y_1, y_2)$ is non-zero. We have

$$D(y_1, y_2) = \begin{vmatrix} y_1 & y_1^* \\ y_1' & y_1^{*'} \end{vmatrix} = \begin{vmatrix} uv & u^*v^* \\ uv' + zv & u^*v^{*'} + z^*v^* \end{vmatrix}.$$

Taking into consideration the fact that $v' = \dfrac{i}{\mu} Qv$, we further obtain

$$D(y_1, y_2) = vv^* \begin{vmatrix} u & u^* \\ u\dfrac{i}{\mu} Q + z & -u^* \dfrac{i}{\mu} Q + z^* \end{vmatrix}$$

$$= \frac{iQ}{\mu} \begin{vmatrix} u & u^* \\ u + O(\mu) & -u^* + O(\mu) \end{vmatrix} = -\frac{2iQ}{\mu} [uu^* + O(\mu)]$$

$$= -\frac{2iQ}{\mu} [\bar{u}^2 + O(\mu)],$$

which already shows that this expression is non-zero.

Thus y_1 and $y_2 = y_1^*$ do indeed constitute a fundamental system of solutions, and Theorem 7.6 is proved.

Remarks. 1. It can be shown that a similar result is valid for the equation $\mu^2 y'' - Q^2(x) y = 0$, with the only difference that in (7.78) we must replace i by 1.

2. The asymptotic formulas (7.78) obtained above lose their meaning if the segment $[a, b]$ contains points on which $Q(x)$ vanishes. Such points are called *turning points*. This terminology comes from quantum mechanics, where certain problems for the Schrödinger equation in the one-dimensional case lead to equations of type (7.74). In the presence of turning points, the asymptotics are constructed in a much more complicated way; the corresponding theory lies beyond the scope of this textbook. The method for constructing the asymptotics of solutions of type (7.74) equations is often called the WKB method in physics, from the names of the scientists Wentzel, Brillouin and Kramers, who developed the corresponding algorithm in connection with quantum mechanics problems (for more details, see [11] Chapter 7).

3. The method was demonstrated on example (7.74). It should, however, be kept in mind that this method can be generalized for singularly perturbed systems of more general form. A detailed analysis of the main ideas of this method, its further development and relationship with the so-called regularization of singularly perturbed problems is contained in the work of S. A. Lomov (see, for example [12]).

4. The averaging method. Consider the equation

$$\frac{dy}{dt} = \mu Y(y, t), \qquad y(0, \mu) = y^0. \tag{7.86}$$

We know from the above that for a sufficiently smooth right-hand side in (7.86) on a certain closed interval $[0, T]$, the solution of problem (7.86) can be represented asymptotically by a polynomial in μ (Theorem 7.1). However, in order to answer a series of questions of mathematical physics, it is necessary to investigate the solution for t in arbitrarily large intervals, for example, for $t \sim 1/\mu$. In this case, the methods described in §1 cannot be applied.

It is natural to relate this case to the class of non-regular perturbed problems. Note that the change of variables $\xi = t\mu$ will give us a problem where ξ varies in a finite interval, but the equation acquires the form

$$\frac{dy}{d\xi} = Y\left(y, \frac{\xi}{\mu}\right).$$

In this form the nonregularity of the perturbation becomes especially noticeable.

In this subsection, we shall describe one more asymptotic method developed for nonregular perturbed systems – the so-called *averaging method*, whose foundations were developed by N. M. Krylov and N. N. Bogolyubov [10]. This method, as will be seen later, is especially convenient to describe nonlinear oscillation processes.

Let us state the rule for constructing the asymptotics given by the averaging method. Introduce the functions

$$\bar{Y}(\eta) = \lim_{T \to \infty} \frac{1}{T} \int_0^T Y(\eta, t)\, dt, \qquad (7.87)$$

which are the *average values* of the right-hand side with respect to the variable t which appears explicitly in them. It is assumed that the variable η is fixed.

Remark. In the case of bounded periodic functions with respect to t (to be definite, we shall assume the period equal to 2π), we have

$$\bar{Y}(\eta) = \frac{1}{2\pi} \int_0^{2\pi} Y(\eta, t)\, dt.$$

Indeed, $T = 2k\pi + \tau$ ($0 \leq \tau \leq 2\pi$) and

$$\lim_{T \to \infty} \frac{1}{T} \int_0^T Y(\eta, t)\, dt = \lim_{k \to \infty} \frac{1}{2k\pi \left(1 + \dfrac{\tau}{2k\pi}\right)} \int_0^{2k\pi + \tau} Y(\eta, t)\, dt$$

$$= \lim_{k \to \infty} \frac{1}{2k\pi} \int_0^{2k\pi} Y(\eta, t)\, dt = \frac{1}{2\pi} \int_0^{2\pi} Y(\eta, t)\, dt.$$

Assume that besides the limiting relations (7.87) we also have the relations

$$\frac{d\bar{Y}}{d\eta} = \lim_{T \to \infty} \frac{1}{T} \int_0^T \frac{\partial Y}{\partial \eta}(\eta, t)\, dt, \qquad (7.88)$$

i.e. the averaged value of the derivative $\dfrac{\partial Y}{\partial \eta}$ equals the derivative of the mean value $\bar{Y}$.

Instead of equation (7.86), let us consider the following equation which is called the averaged equation and is, in principle, simpler than (7.86):

$$\frac{d\eta}{dt} = \mu \bar{Y}(\eta), \qquad \eta(0) = y^0. \qquad (7.89)$$

We have the following

Theorem 7.7. *Suppose*

1°. In the domain $|y|\le b$, $0\le t<\infty$, *the function* $Y(y,t)$ *is continuous and uniformly bounded as is its first order derivative with respect to* y.

2°. For $|y|\le b$, *the averaged value (7.87) exists and we also have (7.88), the passage to the limit in (7.87) and (7.88) being uniform with respect to* $\eta\in[-b,b]$.

3°. The solutions y *and* η *of problems (7.86) and (7.89) exist and range over* $(-b,b)$ *for* $0\le t\le L/\mu$, *where* L *is a constant not depending on* μ. *Then we have the following limiting relation, which is uniform with respect to* $t\in[0,L/\mu]$:

$$\lim_{\mu\to 0}\,(y-\eta)=0. \tag{7.90}$$

Consider the expression

$$u(\eta,t)=\int_0^t [Y(\eta,t)-\bar{Y}(\eta)]dt, \tag{7.91}$$

and its derivative with respect to η

$$\frac{\partial u}{\partial\eta}=\int_0^t \left[\frac{\partial Y}{\partial\eta}(\eta,t)-\frac{\partial\bar{Y}}{\partial\eta}\right]dt. \tag{7.92}$$

Let us prove that, in the domain $|\eta|\le b$, $0\le t\le L/\mu$, we have the relations

$$\mu u=\varepsilon(\mu), \qquad \mu\,\frac{\partial u}{\partial\eta}=\varepsilon(\mu). \tag{7.93}$$

From now on let us agree to denote by $\varepsilon(\mu)$ any expression which is infinitely small together with μ uniformly with respect to the other variables (e.g. η and t) on which this expression depends. In other words, (7.93) means that μu and $\dfrac{\partial u}{\partial\eta}$ tend to zero uniformly with respect to η and t.

Let us prove the first of the relations (7.93). The second one is proved in a similar way. We have

$$\mu u(\eta,t)=\mu t\cdot\frac{1}{t}\int_0^t [Y(\eta,t)-\bar{Y}(\eta)]dt=\mu t V(\eta,t).$$

Suppose $0\le t\le 1/\sqrt{\mu}$. Then $\mu t\le\sqrt{\mu}$, $V(\eta,t)$ is uniformly bounded with respect to $\eta\in[-b,b]$, since the passage to the limit in (7.87) is uniform, i.e. $\mu u(\eta,t)=0(\sqrt{\mu})$. In the case $1/\sqrt{\mu}\le t\le L/\mu$, we have $\mu t\le L$ and, by (7.87), $\forall\delta>0\exists\mu_0(\delta)$ such that $|V(\eta,t)|<\delta/L$ for any $\mu<\mu_0(\delta)$, i.e. $|\mu u|<\delta$ for all $|\eta|\le b$. Therefore, $\mu u(\eta,t)=\varepsilon(\mu)$ for $0\le t\le L/\mu$, which was to be proved.

Now introduce the following function

$$\bar{y}(t,\mu)=\eta(t)+\mu u(\eta(t),t). \tag{7.94}$$

This function satisfies the equation

$$\bar{y}' = \mu Y(\bar{y}, t) + R, \quad \bar{y}(0, \mu) = y^0,$$

where $R = (\eta + \mu u)' - \mu Y(\eta + \mu u, t) = \mu \bar{Y}(\eta) + \mu \dfrac{du}{dt} - \mu Y(\eta + \mu u, t)$. Taking into

consideration the fact that $\dfrac{du}{dt} = \dfrac{\partial u}{\partial \eta} \mu \bar{Y}(\eta) + \dfrac{\partial u}{\partial t}, \dfrac{\partial u}{\partial t} = Y(\eta, t) - \bar{Y}(\eta)$ and $Y(\eta$

$+ \mu u, t) = Y(\eta, t) + \dfrac{\partial Y}{\partial y}(\eta^*, t)\mu u$, we obtain

$$R = \mu^2 \dfrac{\partial u}{\partial \eta} \bar{Y}(\eta) - \mu^2 u \dfrac{\partial Y}{\partial y}(\eta^*, t) = \mu \varepsilon(\mu) = o(\mu).$$

It is easy to verify that $\varDelta = y - \bar{y} \to 0$ (as $\mu \to 0$) uniformly with respect to $t \in [0, L/\mu]$. Indeed, the expression $\varDelta$ satisfies the equation

$$\dfrac{d\varDelta}{dt} = \mu \dfrac{\partial Y}{\partial y}(\bar{y} + \theta\varDelta, t)\varDelta + R. \tag{7.95}$$

Here $0 \le \theta \le 1$, $R = o(\mu)$, $\varDelta(0, \mu) = 0$. By assumption 3° of the theorem, $|\bar{y} + \theta\varDelta| < b$ for $0 \le t \le L/\mu$. Therefore there exists a constant N, not depending on μ, such that $\dfrac{\partial Y}{\partial y}(\bar{y} + \theta\varDelta, t) < N$ and (7.95) implies

$$|\varDelta| = \left| \int_0^t e^{\mu \int_\xi^t \frac{\partial Y}{\partial y}(\bar{y} + \theta\varDelta, t)dt} R d\xi \right| < o(\mu) \dfrac{L}{\mu} e^{NL},$$

which in turn implies the convergence of $\varDelta$ to zero (as $\mu \to 0$) uniformly with respect to $t \in [0, L/\mu]$.

Using this result, the relation (7.94) and the estimate (7.93), we obtain (7.90), so that theorem 7.7 is proved.

Remarks. 1. The statement of Theorem 7.7 remains valid in the case when y in (7.86) is a vector function. Under the assumptions of the theorem, instead of $\dfrac{\partial Y}{\partial y}$ we will naturally have to consider the derivative $\dfrac{\partial Y_i}{\partial y_k}$. In the proof, certain technical complications will arise in the estimate of $\varDelta$, but these may be overcome by using the considerations developed in §4, Chapter 2.

2. Theorem 7.7 was stated and proved under the simplifying assumption that not only the solution $\eta(t, \mu)$ of the averaged system, which was simpler than the given system, is contained in $(-b, b)$, but the solution of the given system $y(t, \mu)$ itself was also located in $(-b, b)$. This last requirement may be eliminated by

imposing a supplementary assumption on $\eta(t,\mu)$: $\eta\in[-b+\delta,b-\delta]$ for $0\le t\le L/\mu$. The inequality $y(t,\mu)<b$ can be obtained as a consequence of this. Indeed, suppose that we have $|y(T,\mu)|=b$ for some $T\le L/\mu$, i.e. integral curve reaches the boundary of the domain. Choose a $T^*<T$ sufficiently close to T so that $y(T^*,\mu)$ differs from $y(T,\mu)$ by no more than $\delta/4$. Since $|y(t,\mu)|<b$ for $0\le t\le T^*$, it follows that, for $0\le t\le T^*$, Theorem 7.7 is valid and, for sufficiently small μ, the value of $\eta(T^*,\mu)$ differs from $y(T^*,\mu)$ by no more than $\delta/4$. But then $\eta(T^*,\mu)$ differs from $y(T,\mu)$, which equals b or $-b$, by no more than $\delta/2$, which contradicts the fact that $\eta\in[-b+\delta,b-\delta]$. This contradiction leads us to the conclusion that $|y(t,\mu)|<b$ for $0\le t\le L/\mu$ and then, as we have shown, (7.90) is valid on the entire interval $[0,L/\mu]$.

Also, we need not have assumed the existence of the solution $y(t,\mu)$ and could have proved its existence in a neighbourhood of $\eta(t)$, as was done, for example, in Theorem 7.2.

3. Here we gave the simplest example of theorem on the averaging method. There is a large amount of literature devoted to the averaging method; in it one may find proofs of various more delicate and complicated theorems and asymptotic approximations of any degree of precision (for a detailed exposition, see [2]).

Example. Consider the Van der Pol equation

$$x''-\mu(1-x^2)x'+x=0 \tag{7.96}$$

which describes a lamb generator based on a triode with an oscillating circuit in the anode chain[4]. Rewrite (7.96) in the form of the system

$$x'=u, \qquad u'=\mu(1-x^2)u-x$$

and take the initial conditions $x(0,\mu)=x^0$, $u(0,\mu)=0$.

Introduce the new variables y_1 and y_2 by putting

$$x=y_1\cos(t+y_2), \qquad u=-y_1\sin(t+y_2).$$

---

[4] See, for example, [14]. In this book, the Van der Pol equation is derived. As for the asymptotics, the authors are interested in the case when μ, instead of being small, is, on the contrary, large. Dividing everything by this large parameter, we obtain an equation in which the small factor multiplies x''. This equation also belongs to the singularly perturbed type, but describes oscillations which are not close to harmonic oscillations as in the case of (7.96), but oscillations of another type, so-called *relaxational oscillations*. E. F. Mishchenko's and N. Kh. Rozov's monograph [14] is devoted to the asymptotic theory of relaxational oscillations for systems of the general form, developed by L. S. Pontrjagin and the authors of the monograph.

With respect to the variables y_1, y_2, we obtain the system

$$y_1' = \mu \left[\frac{y_1}{2} \left(1 - \frac{y_1^2}{4} \right) - \frac{y_1}{2} \cos 2 \, (t + y_2) + \frac{y_1^3}{8} \cos 4 \, (t + y_2) \right],$$

$$y_2' = \mu \left[\frac{1}{2} \left(1 - \frac{y_1^2}{2} \right) \sin 2 \, (t + y_2) - \frac{y_1^2}{8} \sin 4 \, (t + y_2) \right]$$

with the conditions $y_1(0, \mu) = x^0$, $y_2(0, \mu) = 0$; this is a system of type (7.86).
The averaged system for (7.89) is of the form

$$\eta_1' = \mu \frac{\eta_1}{2} \left(1 - \frac{\eta_1^2}{4} \right), \quad \eta_2' = 0; \quad \eta_1(0) = x^0, \quad \eta_2(0) = 0.$$

This gives $\eta_2(t) \equiv 0$. The equation for η_1 separates. If we consider the (t, η_1)-plane, then, just as in Fig. 21 (the given case differs from the one considered there only in the choice of the scale on the t-axis), it is easy to see that for $x^0 < 0$ there exists a solution $\eta_1(t)$ which tends to $\eta_1 = -2$ as $t \to \infty$. The solutions may be written out in quadratures. An approximate solution of equation (7.96) obtained by the averaging method is therefore of the form

$$x = [\eta_1(t, \mu) + \varepsilon(\mu)] \cos \, (t + \varepsilon(\mu)). \tag{7.97}$$

The equation for η_1 has two stationary solutions $\eta_1 = \pm 2$. The corresponding solutions of equation (7.96) are of the form $x = [\pm 2 + \varepsilon(\mu)] \cos \, (t + \varepsilon(\mu))$.

Note that the expansion (7.97) cannot be obtained by developing x into the power series (7.6) with respect to the parameter μ.

Chapter VIII
First Order Partial Differential Equations

First order partial differential equations are traditionally studied in ordinary differential equation courses, because their general solutions can be found by methods developed in the theory of ordinary differential equations.

A partial differential equation of the first order is a relation of the form

$$F\left(x_1, \ldots, x_n, u, \frac{\partial u}{\partial x_1}, \ldots, \frac{\partial u}{\partial x_n}\right) = 0, \tag{8.1}$$

where F is some function of its arguments, while the unknown function is u and depends on the variables $x_1, \ldots, x_n$.

We shall consider here two particular cases of (8.1). They are the so-called *linear equation*

$$\sum_{i=1}^{n} a_i(x_1, \ldots, x_n) \frac{\partial u}{\partial x_i} = 0 \tag{8.2}$$

and the *quasilinear equation*

$$\sum_{i=1}^{n} a_i(x_1, \ldots, x_n, u) \frac{\partial u}{\partial x_i} = a(x_1, \ldots, x_n, u), \tag{8.3}$$

where a_i, a are known functions of their variables.

§1. Linear Equations

We now consider the equation (8.2):

$$a_1(x_1, \ldots, x_n) \frac{\partial u}{\partial x_1} + \ldots + a_n(x_1, \ldots, x_n) \frac{\partial u}{\partial x_n} = 0. \tag{8.4}$$

Suppose $x_1, \ldots, x_n$ vary in a certain domain G and assume that in this domain the functions $a_i(x_1, \ldots, x_n)$ $(i = 1, \ldots, n)$ possess continuous partial derivatives and do not simultaneously vanish, which can be expressed in the form

$$\sum_{i=1}^{n} a_i^2(x_1, \ldots, x_n) \neq 0.$$

A solution of equation (8.4) is any function, possessing partial derivatives with respect to the variables $x_1, \ldots, x_n$, which transforms (8.4) into an identity. Geometrically, the solution may be interpreted as a surface in the space $x_1, \ldots, x_n, u$. This surface will be called an *integral surface*.

1. The two-dimensional case. First we consider the case $n = 2$, in order to clarify the main ideas used in constructing the solution

$$A(x, y) \frac{\partial u}{\partial x} + B(x, y) \frac{\partial u}{\partial y} = 0. \tag{8.5}$$

Rigorous proofs will be presented in the next subsection, which deals with the multidimensional case.

The expression in the left-hand side may be interpreted as the scalar product of vector fields $\{A, B\}$ and grad $u = \left\{ \dfrac{\partial u}{\partial x}, \dfrac{\partial u}{\partial y} \right\}$, so that (8.5) means that the derivative of u in the direction of the vector $\{A, B\}$ equals zero. Denote by Γ the curve in the (x, y)-plane such that the tangent vector to this curve is collinear to $\{A, B\}$. Its parametric representation (the parameter is t) may be obtained from the system of equations

$$\frac{dx}{dt} = A(x, y), \qquad \frac{dy}{dt} = B(x, y). \tag{8.6}$$

The phase trajectories of system (8.6) in the (x, y)-plane, which are the integral curves of the equation

$$\frac{dy}{dx} = \frac{B(x, y)}{A(x, y)} \quad \left(\text{or the equation } \frac{dx}{dy} = \frac{A(x, y)}{B(x, y)} \right), \tag{8.7}$$

are called *characteristics of the partial differential equation* (8.5). Instead of the pair of equations (8.7), one often considers one equation in symmetric form (with respect to x and y)

$$\frac{dx}{A(x, y)} = \frac{dy}{B(x, y)}. \tag{8.8}$$

In view of the conditions imposed on A and B $(A^2 + B^2 \neq 0)$, one and only one characteristic passes through each point of G (see Chapter 2, §2, remark 7).

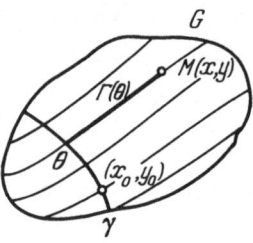

Fig. 22

Suppose $u = u(x, y)$ is the integral surface of equation (8.5). We shall consider it over the characteristic. Then u will be viewed as a compound function of t: $u = u(x(t), y(t))$. It is easy to see that the complete derivative of u with respect to t, by (8.6), will coincide with the left-hand side of (8.5):

$$\frac{du}{dt} = \frac{\partial u}{\partial x}\frac{dx}{dt} + \frac{\partial u}{\partial y}\frac{dy}{dt} = A(x, y)\frac{\partial u}{\partial x} + B(x, y)\frac{\partial u}{\partial y}. \tag{8.9}$$

Therefore, in view of equation (8.5), we have $\dfrac{du}{dt} = 0$, i.e. over the characteristic the u-coordinate of the integral surface has a constant value, i.e. the characteristic is a level line of the integral surface.

Consider some curve γ which does not coincide with the characteristic (Fig. 22). Draw the characteristic through each point $M(x, y)$ of the domain G. Then its intersection of the curve γ uniquely determines this characteristic, i.e. the characteristics form a one-parameter family. For this parameter we can take, for example, the distance θ along the curve γ from some fixed point (x_0, y_0). The position of the point M on each characteristic is determined by the parameter t. As we see from 8.6, t may be chosen up to an arbitrary summand; hence we may assume that the intersection point of each characteristic with the curve γ corresponds to the value $t = t_0$.

Thus, in the domain G, we may assign to every point $M(x, y)$ the pair of numbers (θ, t): θ determines the characteristic passing through M, while t is the value of the parameter on the characteristic corresponding to the point M. Thus we have a one-to-one correspondence $(x, y) \leftrightarrow (\theta, t)$ or analytically

$$x = X(\theta, t), \qquad y = Y(\theta, t)$$

$$\theta = \Theta(x, y) \qquad t = T(x, y). \tag{8.10}$$

In the variables (θ, t), the equation of the characteristic is of the form $\dfrac{d\theta}{dt} = 0$, when $\theta = C$. Thus along the characteristic we have

$$\Theta(x, y) = C. \tag{8.11}$$

The variables θ, t are convenient because, using them, we can very easily obtain a solution of equation (8.5). Denote

$$u(x, y) = u(X(\theta, t), Y(\theta, t)) = v(\theta, t).$$

In view of (8.10), the equation (8.5) in the variables v, θ, t is of the form $\dfrac{\partial v}{\partial t} = 0$ (the solution retains its value along the characteristic). Since $\dfrac{\partial v}{\partial t} = 0$, we see that v depends on θ only, i.e. $v = F(\theta)$, so that

$$u(x, y) = F(\Theta(x, y)), \tag{8.12}$$

where F is an arbitrary function of one variable.

Thus we come to the following conclusion: *the general solution of equation (8.5) is of the form (8.12), where F is an arbitrary function of the argument* $\Theta(x, y)$, *while* $\Theta(x, y)$ *is the left-hand side of (8.11)*.

On the other hand, if we have been able to find some function $\varphi(x, y)$ which is identically constant along every integral curve of equation (8.8), i.e. along each characteristic, then viewed as a function of the variables (θ, t), this function must depend only on θ and not on t, i.e. $\varphi(X(\theta, t), Y(\theta, t)) = \tilde{\varphi}(\theta)$. Since, in (8.12), F was an arbitrary function, the expression for the general solution may be written not only in the form (8.12), but also in the form

$$u(x, y) = F(\varphi(x, y)). \tag{8.13}$$

Indeed, $F(\varphi) = F(\tilde{\varphi}(\theta)) = \tilde{F}(\theta)$, and we again obtain (8.12).

Example 8.1. Suppose in (8.5) the coefficients are constant $A = A_0$, $B = B_0$. From (8.7), find $y - v_0 x = \text{const}$, where $v_0 = B_0/A_0$. Then, according to (8.13), the general solution will be of the form

$$u(x, y) = F(y - v_0 x).$$

The function $F(y - v_0 x)$ is called a running wave of velocity v_0 and profile $F(y)$. It is easy to understand the meaning of this terminology if we interpret x as time, denoting it by t. Let us sketch the profile of the wave at moments t_1 and t_2 (Fig. 23). When $v_0 > 0$ and $t_2 > t_1$, the second profile is shifted to the right without deformation by the shift of value $\Delta y = (t_2 - t_1)v_0$. Indeed, $y + \Delta y - v_0 t_2 = y - v_0 t_1$, i.e. at the points y and $y + \Delta y$ the complete argument of F is the same, while $\Delta y / \Delta t = v_0$, i.e. the velocity of the shift is v_0.

According to (8.12) or (8.13), the general solution of the given partial differential equation depends on an arbitrary function, i.e. contains an even greater degree of freedom than in the case of an ordinary differential equation. Now let us see how a unique solution may be specified from the set (8.12).

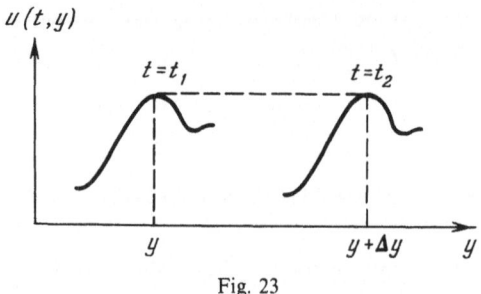

Fig. 23

Suppose that on the curve γ_1, whose equation $x=x(s)$, $y=y(s)$ does not coincide with the characteristic on any interval of positive length, we have the supplementary condition

$$u(x,y)|_{\gamma_1} = \omega(s), \qquad (8.14)$$

where $\omega(s)$ is a given function of the variable s. This problem is known as the initial value problem or the Cauchy problem for equation (8.5).

Since γ_1 is not a characteristic, it follows that $\Theta(x,y)$ changes along γ_1: $\Theta(x,y)|_{\gamma_1} = \xi = \Theta(x(s), y(s))$, i.e. is a function of s. Thus $s = \Omega(\xi)$ while $\omega(s) = \omega[\Omega(\xi)]$. The value of u at the point ξ on the curve γ is $\omega[\Omega(\xi)]$. Now if we construct the characteristic passing through the point $\xi \in \gamma_1$, then its equation will be of the form $\Theta(x,y) = \xi$, while the value of the solution at points of this characteristic equals $\omega[\Omega(\Theta(x,y))]$. But since only one characteristic passes through each point of G, this formula is the expression for the solution at any point of the domain G:

$$u(x,y) = \omega[\Omega(\Theta(x,y))]. \qquad (8.15)$$

The corresponding geometrical picture is presented on Fig. 24.

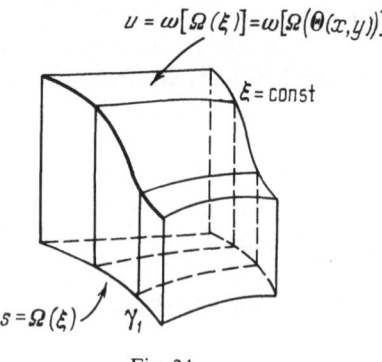

Fig. 24

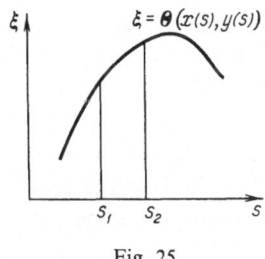

Fig. 25

It is also easy to check directly that formula (8.15) gives the necessary solution: first of all, this is a solution, since it is contained in (8.12), and secondly, $u(x, y)|_{\gamma_1} = \omega[\Omega(\xi)] = \omega(s)$, i.e. the condition (8.14) is satisfied.

Remarks. 1. Instead of $\Theta(x, y)$ we could have used $\varphi(x, y)$ (see (8.13)) in all these arguments.

2. One must keep in mind that, in order to obtain $\Omega(\xi)$, ξ must depend on s along γ_1 monotonically, as, for example, on the segment (s_1, s_2) (Fig. 25). Then $s = \Omega(\xi)$ is uniquely determined.

Example 8.2. Consider the equation from example (8.1), and for (8.14) choose the condition

$$u(t, 0) = \mu(t). \tag{8.16}$$

Here γ_1 is the line $y = 0$ and, for the parameter s, we use t. In this case

$$\varphi(t, y) = y - v_0 t, \; \varphi(t, y)|_{\gamma_1} = -v_0 t = \xi, t = \Omega(\xi) = -\xi/v_0.$$

Therefore the required solution is

$$u(t, y) = \mu\left(\frac{\varphi(t, y)}{-v_0}\right) = \mu\left(t - \frac{y}{v_0}\right).$$

Thus the solution is a running wave whose profile is uniquely determined by the given function $\mu(t)$.

Now suppose γ_1 coincides with some characteristic. Here different cases may arise.

a) $u(x, y)|_{\gamma_1} = \text{const} = u_0$. Then, obviously, the solution is not determined uniquely, since a solution of this problem will be given by any function $u(x, y) = f(\Theta(x, y))$ as long as $f(\Theta(x, y)|_{\gamma_1}) = u_0$. For example, we can obtain

$$f(\Theta(x, y)) = F(\Theta(x, y)) - F(\Theta(x, y)|_{\gamma_1}) + u_0$$

(recall that $F(\Theta(x, y)|\gamma_1) = \text{const}$) where F is now an arbitrary function.

b) $u(x,y)|_{\gamma_1} \neq$ const. In this case the solution of the problem does not exist, since any solution of equation (8.5) is constant along the characteristic and therefore the requirement imposed on γ_1 cannot be satisfied.

2. The multidimensional case. Now let us consider the general case of (8.4) under the assumptions stated at the beginning of § 2.

To the equation (8.4) let us assign the system (compare with (8.6))

$$\frac{dx_i}{dt} = a_i(x_1, \ldots, x_n) \quad (i = 1, \ldots, n) \tag{8.17}$$

and the system for the phase trajectories (compare with (8.8))

$$\frac{dx_1}{a_1} = \ldots = \frac{dx_n}{a_n}. \tag{8.18}$$

The integral curves of the system of equations (8.18) are called *characteristics* of the partial differential equation (8.4). In view of the conditions imposed on the coefficients a_i, the existence and uniqueness theorem holds for the system (8.18), so that there is one and only one characteristic passing through every point of the n-dimensional domain G.

Theorem 8.1. *Along the characteristics, the solution $u(x_1, \ldots, x_n)$ of equation (8.4) has a constant value.*

Indeed, (compare with (8.9))

$$\frac{du}{dt} = \sum_{i=1}^{n} \frac{\partial u}{\partial x_i} \frac{dx_i}{dt} = \sum_{i=1}^{n} \frac{\partial u}{\partial x_i} a_i = 0, \tag{8.19}$$

i.e. $u = $ const.

The idea of the construction of the general solution of equation (8.4) is a natural generalization of what we had in the case $n = 2$. The domain G is covered by characteristics which constitute a family depending on $n-1$ parameters $\theta_1, \ldots, \theta_{n-1}$. To each point $(x_1, \ldots, x_n)$ of the domain G we can assign a system of values $\theta_1, \ldots, \theta_{n-1}, t$ so that $\theta_i = \Theta_i(x_1, \ldots, x_n)$, $t = T(x_1, \ldots, x_n)$. The choice of $\theta_1, \ldots, \theta_{n-1}$ specifies one of the characteristics from the family, while t is a parameter which determines the point on this characteristic. We have $u(x_1, \ldots, x_n) = v(\theta_1, \ldots, \theta_{n-1}, t)$. In the variables $\theta_1, \ldots, \theta_{n-1}, t$, equation (8.4) becomes $\dfrac{\partial v}{\partial t} = 0$. Thus $v = F(\theta_1, \ldots, \theta_{n-1})$ and therefore [compare with (8.12)] we have

$$u(x_1, \ldots, x_n) = F(\Theta_1(x_1, \ldots, x_n), \ldots, \Theta_{n-1}(x_1, \ldots, x_n)). \tag{8.20}$$

In order to establish this formula rigorously, we shall need the notion of the first integral of the system of equations (8.18). Now we shall assume that

$a_n(x_1,\ldots,x_n) \neq 0$ in G. Then (8.18) may be rewritten in the form of a normal system

$$\frac{dx_i}{dx_n} = \frac{a_i}{a_n} \quad (i=1,\ldots,n-1). \tag{8.21}$$

and for $\theta_1,\ldots,\theta_{n-1}$ we may take the initial values $x_1^0,\ldots,x_{n-1}^0$; x_n^0; x_n will play the role of t. The family of solutions of the system (8.21) is of the form

$$x_i = X_i(x_n, x_1^0,\ldots,x_{n-1}^0, x_n^0) \quad (i=1,\ldots,n-1). \tag{8.22}$$

Here X_i expresses the relationship between two points on the integral curve: the initial point and the generic point. Interchanging them, we obtain

$$x_i^0 = X_i(x_n^0, x_1,\ldots,x_{n-1}, x_n) \quad (i=1,\ldots,n-1), \tag{8.23}$$

where the X_i are the same functions as in (8.22), since they express the same relationship.

Remark. The fact that we have the same functions in the right-hand side of (8.22) and (8.23) can conveniently be demonstrated with the example of a linear system with independent variable x_n and unknown vector functions x with components $x_1,\ldots,x_{n-1}$. The formula (8.22) is of the form (see 3.84) $x = W(x_n) W^{-1}(x_n^0) x_n^0 \equiv \mathscr{K}(x_n, x_n^0) x^0$. Solving this with respect to x^0, we obtain (8.23), i.e.

$$x^0 = W(x_n^0) W^{-1}(x_n) x = \mathscr{K}(x_n^0, x_n) x.$$

The functions $X_i(x_1,\ldots,x_n)$ (x_n^0 is fixed in (8.23)) may be used for Θ_i in the construction of the general solution (8.20). Note also that since the formulas (8.22) and (8.23) are inverse to each other, it follows that

$$\frac{D(X_1,\ldots,X_{n-1})}{D(x_1,\ldots,x_{n-1})} \neq 0 \text{ in } G.$$

Definition. *A first integral* of the system (8.18) is a function $\varphi(x_1,\ldots,x_n)$ which is identically constant when the point $(x_1,\ldots,x_n)$ moves along an integral curve of the system (8.18). (Sometimes it is not the function φ, but the relation $\varphi =$ const which is referred to as the first integral.)

Obviously, the functions $X_i(x_1,\ldots,x_n)$ in formulas (8.23) are first integrals of system (8.21), since when we substitute (8.22) in the right hand sides of (8.23) we obtain x_i^0 identically. However, other functions may also be first integrals and, which is especially convenient for practical solutions, first integrals may often be obtained by the method of integrable combinations (in order to obtain (8.23), one must solve the initial value problem, and this is less convenient).

Example 8.3. $\dfrac{dx_1}{dx_3} = x_2$, $\dfrac{dx_2}{dx_3} = x_1$. Adding these equations, we obtain $\dfrac{d}{dx_3}(x_1 + x_2) = (x_1 + x_2)$. Hence $x_1 + x_2 = Ce^{x_3}$ and the first integral will be

$$\varphi_1 = (x_1 + x_2)e^{-x_3}.$$

Subtracting the equations, we obtain $\dfrac{d}{dx_3}(x_1 - x_2) = -(x_1 - x_2)$, so that we find another first integral

$$\varphi_2 = (x_1 - x_2)e^{x_3}.$$

Suppose we have found $n-1$ first integrals $\varphi_i(x_1, \ldots, x_n)$ $(i = 1, \ldots, n-1)$ of system (8.18) and

$$\frac{D(\varphi_1, \ldots, \varphi_{n-1})}{D(x_1, \ldots, x_{n-1})} \neq 0 \text{ in } G. \tag{8.24}$$

Integrals $\varphi_1, \ldots, \varphi_{n-1}$ satisfying (8.24) are called independent ((8.23) is an example of a system of $n-1$ independent first integrals).

Theorem 8.2. *Any solution $\psi(x_1, \ldots, x_n)$ of equation (8.4) is a first integral of system (8.18) and, conversely, every first integral $\varphi(x_1, \ldots, x_n)$ of system (8.18) is a solution of equation (8.4).*

The first statement was actually proved earlier (see Theorem 8.1 and the chain of identities (8.19), where $u = \psi$).

To prove the converse statement, replace u by φ in (8.19) and begin with the identity $\dfrac{d\varphi}{dt} \equiv 0$ obtaining the final identity $\sum\limits_{i=1}^{n} a_i \dfrac{\partial \varphi}{\partial x_i} \equiv 0$. This last is an identity only with respect to the variable t, i.e. along characteristics. But in order to claim that $\varphi(x_1, \ldots, x_n)$ is a solution of equation (8.4), it is necessary that this identity be an identity with respect to all the variables $x_1, \ldots, x_n$. However, since one characteristic passes through each point of G, the fact that the expression under consideration vanishes identically along each characteristic implies that it is identically equal to zero in all of G.

Theorem 8.3. *Every solution $u = \psi(x_1, \ldots, x_n)$ of equation (8.4) can be represented in the form*

$$u = \Psi(\varphi_1(x_1, \ldots, x_n), \ldots, \varphi_{n-1}(x_1, \ldots, x_n)), \tag{8.25}$$

where $\Psi(\varphi_1, \ldots, \varphi_{n-1})$ is some differentiable function of its arguments $\varphi_1, \ldots, \varphi_{n-1}$, while $\varphi_i(x_1, \ldots, x_n)$ $(i = 1, \ldots, n-1)$ are first integrals of the system (8.18) satisfying condition (8.24).

Proof. ψ, as well as the φ_i (according to Theorem 8.2) are solutions of (8.4), i.e.

$$a_1 \frac{\partial \psi}{\partial x_1} + \ldots + a_n \frac{\partial \psi}{\partial x_n} = 0,$$

$$a_1 \frac{\partial \varphi_1}{\partial x_1} + \ldots + a_n \frac{\partial \varphi_1}{\partial x_n} = 0.$$

$$. \quad . \quad . \quad . \quad . \quad . \quad . \quad . \quad . \quad . \quad . \quad .$$

$$a_1 \frac{\partial \varphi_{n-1}}{\partial x_1} + \ldots + a_n \frac{\partial \varphi_{n-1}}{\partial x_n} = 0.$$

At each point of the domain G these relations may be considered as systems of linear algebraic equations with respect to $a_1, \ldots, a_n$. By assumption, $\sum_{i=1}^{n} a_i^2 \neq 0$, i.e. there is a non-trivial solution. Therefore, the determinant of this system equals zero in the entire domain G:

$$\frac{D(\psi, \varphi_1, \ldots, \varphi_{n-1})}{D(x_1, \ldots, x_n)} \equiv 0.$$

Hence from well-known theorem in calculus (see [3], p. 155) there must be a functional dependence between $\psi, \varphi_1, \ldots, \varphi_{n-1}$ and, by condition (8.24), this dependence may be represented solved with respect to ψ, in the form

$$\psi = \Psi(\varphi_1, \ldots, \varphi_{n-1}).$$

as required.

Remarks. 1. The theorem proved above establishes formula (8.20) given earlier.

2. Formula (8.25) for an arbitrary differentiable Ψ, possesses the same property and it contains, according to Theorem 8.3, any solution of equation (8.4). On the other hand, it is easy to check directly that for any differentiable Ψ the function u from (8.25) satisfies equation (8.4), so that formula (8.25) is in fact the general solution of equation (8.4).

Now let us state the supplementary condition which allows us to specify one solution from the set (8.25). To do this, we must choose in the domain G a manifold of dimension $n-1$ and, on this manifold, choose the value of the required solution (in the case $n=2$ in subsection 1 we considered the curve γ_1 in the role of this manifold). Suppose this manifold (which shall also be denoted by γ_1) is given parametrically (in terms of the parameters $s_1, \ldots, s_{n-1}$) in the form $x_i = \omega_i(s_1, \ldots, s_{n-1})$ $(i=1, \ldots, n)$ and the required solution on it is also given as a function of the parameters $s_1, \ldots, s_{n-1}$:

$$u|_{\gamma_1} = \omega(s_1, \ldots, s_{n-1}) \tag{8.26}$$

(the initial value or Cauchy problem).

Suppose we know $n-1$ independent first integrals φ_i. We have

$$\varphi_i|_{\gamma_1} = \varphi_i(\omega_1(s_1,\ldots,s_{n-1}),\ldots,\omega_n(s_1,\ldots,s_{n-1})).$$

Denote $\xi_i = \varphi_i(\omega_1(s_1,\ldots,s_{n-1}),\ldots,\omega_n(s_1,\ldots,s_{n-1}))$ $(i=1,\ldots,n-1)$. Assume that this system of $n-1$ equations with $n-1$ unknowns $s_1,\ldots,s_{n-1}$ can be solved with respect to $s_1,\ldots,s_{n-1}$, so that $s_i = \Omega_i(\xi_1,\ldots,\xi_{n-1})$ $(i=1,\ldots,n-1)$. Then the solution of the problem considered will be

$$u(x_1,\ldots,x_n) = \omega[\Omega_1(\varphi_1(x_1,\ldots,x_n),\ldots,\varphi_{n-1}(x_1,\ldots,x_n)),\ldots$$

$$\ldots\Omega_{n-1}(\varphi_1(x_1,\ldots,x_n),\ldots,\varphi_{n-1}(x_1,\ldots,x_n))]. \quad (8.27)$$

Indeed, this expression is the solution of equation (8.4), since it is contained in formula (8.25) (see the remark to Theorem 8.3). Also, taking into consideration the fact that $\varphi_i|_{\gamma_1} = \xi_i$, we obtain

$$u|_{\gamma_1} = \omega[\Omega_1(\xi_1,\ldots,\xi_{n-1}),\ldots,\Omega_{n-1}(\xi_1,\ldots,\xi_{n-1})] = \omega(s_1,\ldots,s_{n-1}),$$

i.e. condition (8.26) is satisfied.

Questions of uniqueness or non-uniqueness of the solution of problem (8.26) will not be considered in general form. For $n=2$ this was done earlier in subsection 1.

§2. Quasilinear Equations

We now consider the equation (8.3):

$$\sum_{i=1}^{n} a_i(x_1,\ldots,x_n,u)\frac{\partial u}{\partial x_i} = a(x_1,\ldots,x_n,u). \quad (8.28)$$

We shall assume that a_i $(i=1,\ldots,n)$ and the a are differentiable functions of their variables $x_1,\ldots,x_n,u$ in some domain D of $(n+1)$-dimensional space of the variables $x_1,\ldots,x_n,u$.

Any function of the arguments $x_1,\ldots,x_n$ possessing partial derivatives with respect to these arguments and transforming the equation (8.28) in an identity will be called a solution of equation (8.28). As in the case of linear equations, this solution may be interpreted geometrically as a surface in the space of the variables $x_1,\ldots,x_n,u$.

1. The general solution and the initial value problem. To equation (8.28) let us assign the following *linear* equation:

$$\sum_{i=1}^{n} a_i(x_1,\ldots,u)\frac{\partial v}{\partial x_i} + a(x_1,\ldots,u)\frac{\partial v}{\partial u} = 0. \quad (8.29)$$

Theorem 8.4. *Suppose* $v = V(x_1, \ldots, x_n, u)$ *is a solution of equation (8.29). Suppose the equation* $V(x_1, \ldots, x_n, u) = 0$ *determines a certain differentiable function* $u = \varphi(x_1, \ldots, x_n)$ *in the domain G of the variables* $x_1, \ldots, x_n$ *and assume that* $\left. \dfrac{\partial V}{\partial u} \right|_{u=\varphi} \neq 0$ *in G. Then* $u = \varphi(x_1, \ldots, x_n)$ *is a solution of equation (8.3).*

Indeed, according to the theorem on implicit functions,

$$\frac{\partial \varphi}{\partial x_i} = -\frac{\partial V}{\partial x_i} \bigg/ \frac{\partial V}{\partial u}$$

so that the left-hand side of equation (8.28) equals

$$-\sum_{i=1}^{n} a_i \frac{\partial V}{\partial x_i} \bigg/ \frac{\partial V}{\partial u}.$$

and this, in view of (8.29), is $a(x_1, \ldots, \varphi)$, i.e. equation (8.28) is satisfied.

The theorem we have just proved and the results of the previous section yield the following method for constructing solutions of equation (8.28). We must write out system of equations determining the characteristics of the linear equation (8.29)

$$\frac{dx_1}{a_1} = \ldots = \frac{dx_n}{a_n} = \frac{du}{a}. \tag{8.30}$$

The integral curves of system (8.30), i.e. the characteristics of the linear equation (8.29) will be called *characteristics of the quasilinear equation* (8.28). If equation (8.29) satisfies the conditions imposed on the linear equation in §2, then these characteristics fill up a domain D in the space of variables $x_1, \ldots, x_n, u$ so that one and only one characteristic passes through each point of D. Further, according to formula (8.25), we construct the general solution of equation (8.29) (but now there will be n independent integrals and they will be functions of $x_1, \ldots, x_n, u$):

$$v = \Psi(\varphi_1(x_1, \ldots, x_n, u), \ldots, \varphi_n(x_1, \ldots, x_n, u)). \tag{8.31}$$

Then, setting $v = 0$, we obtain an equation for determining the set of solutions of equation (8.28):

$$\Psi(\varphi_1(x_1, \ldots, x_n, u), \ldots, \varphi_n(x_1, \ldots, x_n, u)) = 0. \tag{8.32}$$

Remark. As was shown in §2, under sufficiently general assumptions formula (8.31) contains all the solutions of equation (8.29). Can we say that formula (8.32) contains all the solutions of equation (8.28)? Let us analyze the proof of theorem 8.4. When we verified the identity (8.28), we used the identity (8.29) and it was sufficient for us that (8.29) be an identity with respect to $x_1, \ldots, x_n$ (then we had $u = \varphi(x_1, \ldots, x_n)$). However, in constructing $V(x_1, \ldots, x_n, u)$ *one requires more, namely (8.29) must be an identity in* $x_1, \ldots, x_n, u$. Hence *a priori* we cannot exclude the possibility that there are solutions of (8.28) for which (8.29) is not satisfied identically with respect to $x_1, \ldots, x_n, u$, but only for $u = \varphi(x_1, \ldots, x_n)$. Such solutions in general are not contained in formula (8.32). They are called special solutions. One can show that a special solution is an exceptional case, and in our further considerations we shall not take them into account.

Unlike the linear case, in the quasilinear case the characteristics are not located in the space $x_1, \ldots, x_n$, but in the space $x_1, \ldots, x_n, u$ and therefore the geometric meaning of the characteristics is different. Let us prove the following fact.

Theorem 8.5. *Any integral surface* $u = f(x_1, \ldots, x_n)$ *consists of characteristics in the sense that a characteristic passes through each point of the surface and lies entirely on it.*

Proof. Consider the system of equations

$$\frac{dx_1}{a_1(x_1, \ldots, x_n, f(x_1, \ldots, x_n))} = \ldots = \frac{dx_n}{a_n(x_1, \ldots, x_n, f(x_1, \ldots, x_n))}, \quad (8.33)$$

$$\frac{dx_i}{dt} = a_i(x_1, \ldots, x_n, f(x_1, \ldots, x_n)), \quad (8.34)$$

which determines curves in the space $x_1, \ldots, x_n$. Choose one of them: $x_i = x_i(t)$. The curve which corresponds to it in the space $x_1, \ldots, x_n, u$ will be

$$x_i = x_i(t), \quad u = f(x_1(t), \ldots, x_n(t)),$$

and lies, by construction, on the integral surface $u = f(x_1, \ldots, x_n)$. Let us prove that this curve is a characteristic. Indeed, since we have (8.33), in order to prove that equation (8.30) is satisfied, it is sufficient to check that $\dfrac{du}{dt} = a$. But

$$\frac{du}{dt} = \sum_{i=1}^{n} \frac{\partial f}{\partial x_i} \frac{dx_i}{dt} = \sum_{i=1}^{n} \frac{\partial f}{\partial x_i} a_i = a$$

and by (8.28) this concludes the proof.

Suppose we do not know the integral surface, but know the characteristics; then, if we are capable of pasting them together into a characteristic smooth surface, this surface will be the integral surface, since the vector $\{a_1, \ldots, a_n, a\}$ tangent to the characteristic will also be tangent to the surface and therefore will be orthogonal to the vector $\left\{\dfrac{\partial f}{\partial x_1}, \ldots, \dfrac{\partial f}{\partial x_n}, -1\right\}$ normal to the surface, while the orthogonality condition precisely means that equation (8.28) is satisfied.

These geometric considerations lead us to the following interpretation of formula (8.32). The system of equations (8.33) gives an n-parametric family of characteristics, which can, by using n first integrals, be represented in the form

$$\varphi_i(x_1,\ldots,x_n,u)=\theta_i \quad (i=1,\ldots,n). \tag{8.35}$$

This family of characteristics fills up the whole of domain D. Our problem is to choose an $(n-1)$-dimensional subset of this set constituting the integral surface. To do this, it is sufficient to impose a relation on the parameters $\theta_1,\ldots,\theta_n$, for example by taking an arbitrary sufficiently smooth relation of the form

$$\Psi(\theta_1,\ldots,\theta_n)=0. \tag{8.36}$$

Substituting (8.35) here, we obtain (8.32).

These same geometric considerations enable us to propose the following procedure for solving the problem with supplementary conditions (the Cauchy problem) which is stated as follows: Construct an integral surface through the $(n-1)$-dimensional manifold Γ_1 in the space $x_1,\ldots,x_n$. This manifold may be written parametrically in the form

$$x_i=\omega_i(s_1,\ldots,s_{n-1}) \quad (i=1,\ldots,n),$$
$$u=\omega(s_1,\ldots,s_{n-1}). \tag{8.37}$$

Thus the problem is essentially the same as in the case of a linear equation.

Now we must impose the relation (8,36), not in an arbitrary way, but by using (8.37). To do this, substitute (8.37) into (8.35):

$$\varphi_i(\omega_1(s_1,\ldots,s_{n-1}),\ldots,\omega(s_1,\ldots,s_{n-1}))=\theta_i \quad (i=1,\ldots,n)$$

and, excluding $s_1,\ldots,s_{n-1}$ from this, obtain the necessary relation (8.36), in which Ψ will now in general be a specific function, while (8.32) gives the solution of problem (8.37).

Without analyzing the special traits that can appear in the general case when this procedure is carried out, let us study it in the three-dimensional case (similarly to the way it was done in §2); we then have the equation

$$A(x,y,u)\frac{\partial u}{\partial x}+B(x,y,u)\frac{\partial u}{\partial y}=C(x,y,u). \tag{8.38}$$

Choose Γ_1 (in the three-dimensional case it is a curve):

$$x=X(s), \quad y=Y(s), \quad u=U(s). \tag{8.39}$$

Writing out the system (8.30), which is of the form,

$$\frac{dx}{A}=\frac{dy}{B}=\frac{du}{C},$$

we can find two first integrals $\varphi_i(x, y, u)$ $(i=1, 2)$. Substituting into (8.39), we obtain

$$\varphi_i(X(s), Y(s), U(s)) = \theta_i \quad (i=1, 2). \tag{8.40}$$

Excluding the unknown s from these two equations, we will find a relation

$$\Psi(\theta_1, \theta_2) = 0. \tag{8.41}$$

Thus the equation of the required surface will be

$$\Psi(\varphi_1(x, y, u), \varphi_2(x, y, u)) = 0. \tag{8.42}$$

The geometrical meaning of the procedure is quite simple: from every point of the given curve Γ_1 we construct its characteristic, and all these characteristics taken together form the required integral surface (Fig. 26).

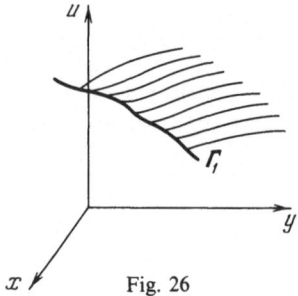

Fig. 26

Instead of the representation (8.42) for the integral surface, which is written in the form of a direct relation between x, y, u, it is sometimes convenient to have a parametric presentation (if for the parameters we take s and, say, x). Suppose, for example, $C(x, y, u) = 0$. Then one of the first integrals equals u, the other $\varphi_2(x, y, u)$ and the required surface may be written in the form

$$x = x, \ u = U(s), \ y = y(x, s), \tag{8.43}$$

where $y(x, s)$ is determined implicitly by the equation

$$\varphi_2(x, y, U(s)) = \varphi_2(X(s), Y(s), U(s)).$$

This last equation gives the projection on the (x, y)-plane of the characteristic which passes through the point s of the given curve Γ_1, while $u = U(s)$ is the value of the u-coordinate above this projection.

In carrying out the procedure which leads to (8.42), we may encounter a special case, namely the one when the left-hand sides of both equations (8.40) become constants and (8.40) will then be of the form

$$\theta_i = c_i = \text{const.} \tag{8.44}$$

In this case, the relation (8.41) also holds, but it is then more arbitrary; namely, Ψ can be any function, as long as $\Psi(c_1, c_2) = 0$. For example, we can put

$$\Psi(\theta_1, \theta_2) = \Phi(\theta_1, \theta_2) - \Phi(c_1, c_2),$$

where Φ is now an arbitrary function of two variables.

The relation (8.44) means that Γ_1 is a characteristic and therefore we can construct an infinite number of integral surfaces passing through a characteristic.

Example 8.4. Consider the transfer equation (1.30) from Chapter 1 in which v is constant,

$$\frac{\partial u}{\partial t} + v\,\frac{\partial u}{\partial x} + c(x, t)u = 0$$

with condition (1.32), where we put

$$u(0, t) = u_0(t).$$

In this case, (8.30) will be of the form

$$\frac{dt}{1} = \frac{dx}{v} = -\frac{du}{c(x, t)u}.$$

Here we can immediately find one of the first integrals: $\varphi_1 = x - vt$. Further we have

$$\frac{du}{dt} = -c(x, t)u = -c(\varphi_1 + vt, t)u = b(t, \varphi_1)u,$$

so that $u = \text{const } e^{b_1(t,\varphi_1)}$, where b_1 is one of the indefinite integrals of b (recall that $\varphi_1 = \text{const}$). Thus the other first integral will be

$$\varphi_2 = ue^{-b_1(t, x - vt)}.$$

For $x = 0$ we have (8.40):

$$\theta_1 = -vt, \qquad \theta_2 = u_0(t)e^{-b_1(t, -vt)}.$$

Thus we find the relation (8.41):

$$\theta_2 = u_0 \left(-\frac{\theta_1}{v} \right) e^{-b_1 \left(-\frac{\theta_1}{v}, \theta_1 \right)},$$

and formula (8.42) yields

$$u e^{-b_1 (t, x - vt)} = u_0 \left(t - \frac{x}{v} \right) e^{-b_1 \left(t - \frac{x}{v}, x - vt \right)}.$$

We can also find the expression for $u(x, t)$ explicitly:

$$u = u_0 \left(t - \frac{x}{v} \right) e^{b_1 (t, x - vt) - b_1 \left(t - \frac{x}{v}, x - vt \right)}.$$

Example 8.5. A linear equation may be viewed as a quasilinear one. Then $a = 0$ and certainly one of the first integrals is u. This corresponds to the fact, established in §1, that along characteristics we have $u = \text{const}$ (Theorem 8.1).

Consider the equation

$$y \frac{\partial u}{\partial x} - x \frac{\partial u}{\partial y} = 0 \tag{8.45}$$

and let us solve it in the "quasilinear way". The corresponding system (8.32) yields two first integrals $\varphi_1 = u$, $\varphi_2 = x^2 + y^2$. Formula (8.32) gives $\Psi(u, x^2 + y^2) = 0$ so that $u = \psi(x^2 + y^2)$. Thus the solution of equation (8.45) is an arbitrary smooth surface of revolution.

Now consider some different supplementary conditions:

A. Γ_1 is the line $x = s$, $y = s$, $u = s$. The method described above gives $\theta_1 = s$, $\theta_2 = 2s^2 \Rightarrow \theta_2 = 2\theta_1^2$, and the solution of the problem will be $x^2 + y^2 = 2u^2$, which is a cone.

B. Γ_1 is the circle: $x = \cos t$, $y = \sin t$, $u = 1$. Thus Γ_1 is the characteristic equation (8.45) viewed as a quasilinear equation. In this case, we have $\theta_1 = 1$, $\theta_2 = 1$ and the solution will be a surface of the form $\Phi(u, x^2 + y^2) - \Phi(1, 1) = 0$. It is geometrically clear that an infinite number of surfaces of revolution pass through the given circle.

C. Viewing (8.45) as a linear equation, we can display the case described at the end of subsection 1 in §2, when the solution of the Cauchy problem does not exist. Suppose γ_1 is given in the form $x = \cos t$, $y = \sin t$, while $u|_{\gamma_1} = t$. Any surface of revolution has a constant coordinate on a given curve γ_1, and therefore the condition $u|_{\gamma_1} = t$ cannot be satisfied.

2. The notion of discontinuous solution. Shock waves. In the present subsection we consider a phenomenon typical of quasilinear equations, which was not observed in the linear case. We shall demonstrate this phenomenon on the simple example of a quasilinear equation in three-dimensional space (in the variables t, x, u), putting $C(t, x, u) = 0$. This case was already described in the previous subsection and was used to demonstrate the parametric form of the expression (8.43) of the surface (8.42). In this case, we have first integrals u and $\varphi_2(t, x, u)$, while the family of characteristics (8.35) is of the form $u = \theta_1$, $\varphi_2(t, x, u) = \theta_2$. In general the projections of these characteristics on the (t, x)-plane can intersect (unlike Fig. 21, which concerns linear equations). This leads us to the following result, which seems paradoxical at first.

Consider the following example

$$\frac{\partial u}{\partial t} + u^2 \frac{\partial u}{\partial x} = 0.$$

Choose the supplementary condition $u(0, x) = u_0(x)$. In our case, the curve Γ_1 is $t = 0$, $x = s$, $u = u_0(s)$. The first integrals will be u and $x - u^2 t$. We can obtain a representation for the required integral surface by using formula (8.43). It will be of the form

$$t = t, \quad x = s + u_0^2(s)t, \quad u = u_0(s). \tag{8.46}$$

The equation $x = s + u_0^2(s)t$ describes the projection (on the (t, x)-plane) of the characteristic which passages through the point s of the curve Γ_1, while $u_0(s)$ is the value of u over this projection.

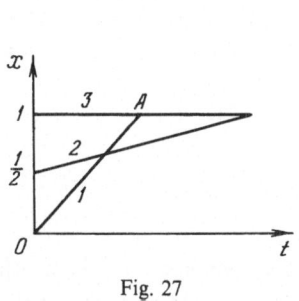

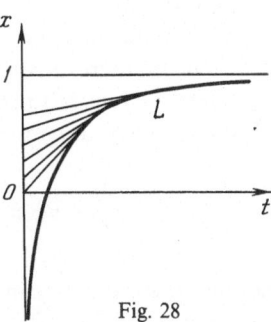

Fig. 27   Fig. 28

Let us choose, specifically, $u_0(s) = 1 - s$. Consider the characteristic passing through the point Γ_1 corresponding to $s = 0$. Its projection is $x = t$ (the line 1 on Fig. 27) to the point $s = 1/2$ corresponds the characteristic with projection $x = \frac{1}{2} + \frac{1}{4} t$ (the line 2) and to the point $s = 1$ – the characteristic with projection $x = 1$ (the line 3). Over line 1, we have $u = u_0(s) = 1$, over line 2, we have $u = 1/2$ and over line 3, we have $u = 0$. But lines 1, 2 and 3 intersect and we see that, for example, at the point A the values of the u-coordinate of the required integral surface must be simultaneously equal to zero and to one (!).

In order to explain this apparent paradox, let us find the solution of the given problem in explicit form. We have $u = 1 - s$, $x = s + (1 - s)^2 t$; hence, excluding s, we obtain

$$u = \frac{1 - \sqrt{1 - 4t(1 - x)}}{2t} \tag{8.47}$$

(the root is taken with the minus sign in order to satisfy the condition $u(0, x) = 1 - x$).

It is obvious from the expression (8.47) that the solution is defined to the left of the hyperbola $4t(1 - x) = 1$ (Fig. 28 the curve L). At the points of the hyperbola L, the expression under the square root sign vanishes, and if t continues to increase, it becomes negative. Thus at the points of the hyperbola the solutions cease to exist. Note that in this case $\dfrac{\partial u}{\partial x} \to 0$. It is easy to check that on the line L the condition $\dfrac{\partial V}{\partial u} \neq 0$, which appeared in Theorem 8.4, fails to hold.

To the left of the hyperbola, the projections of the characteristics do not intersect and there is no "multivaluedness" in the solution. The line $x = t$ "does not reach" the line $x = 1$, so that the line $x = t$ must first intersect the hyperbola and, on the latter, the solution will cease to exist.

Remark. The hyperbola L is the place where infinitely close projections intersect; in the given case it is the envelope of the family $x = s + (1 - s)^2 t$. It can be found by using the s-discriminant curve of this family.

The situation when the solution exists only in a bounded domain can be observed in the case of ordinary differential equations as well, but the case considered here is interesting insofar as it is directly related to some of the physical problems discussed previously.

Recall the problem in subsection 4 §2 of Chapter 1

$$\frac{\partial u}{\partial t} + p(u) \frac{\partial u}{\partial x} = 0, \quad u(x, 0) = u_0(x).$$

The function $p(u)$, which can be obtained experimentally, is of the type illustrated on Fig. 29. Thus the physically interesting case does not differ too

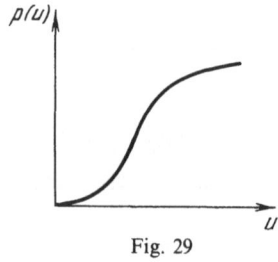

Fig. 29

much from the example $p=u^2$ which was just studied. It is natural to expect that in the physical problem considered the solution will cease to exist on some line L and the derivative with respect to x will become infinite, as it was in the previous example. But the physical process actually continues beyond the lines L, i.e. for greater t. The following question naturally arises: how can we find the solution corresponding to the given physical process which would be valid beyond the line L?

For the mathematical problem considered, the solution may be written in a form similar to (8.46):

$$t=t, \quad x=s+p[u_0(s)]t, \quad u=u_0(s). \tag{8.48}$$

The line L may be found, just as in the example, by using the s-discriminant curve, i.e. by excluding s from the system

$$x=s+p[u_0(s)]t, \quad 0=1+p'[u_0(s)]u_0'(s)t.$$

The curve L is conveniently presented in parametric form (again in terms of s) as follows:

$$t=-\frac{1}{p'[u_0(s)]u_0'(s)}, \quad x=s-\frac{p[u_0(s)]}{p'[u_0(s)]u_0'(s)}.$$

In the given case, as in the example, the projections of the characteristics are straight lines, the slope of the line which passes through the point $t=0$, $x=s$ is $p[u_0(s)]$. Suppose $u_0(s)$ decreases when s increases. Since $p(u)$ increases when u increases (see Fig. 29), the slope of the characteristic decreases when s increases, i.e. the characteristics "converge" to the line L, as shown on Fig. 28 for the example $p=u^2$.

If we observe the process for a fixed $x=x_0$, then, when t approaches t_0 (the point (t_0, x_0) being on L), the derivative $\dfrac{\partial u}{\partial x}$ increases, i.e. a strong change of density is observed. Observing the actual physical phenomenon, we can indeed register, at a definite moment in time, a strong change of density: a so-called *shock wave* passes through the point x_0 at the moment t_0. Then the process continues sufficiently smoothly until a new shock wave is formed, but it can happen that no such new shock wave will arise.

How can we describe the process after the passage of the shock wave? The solution defined as a differentiable function satisfying equation (8.38) enables us to describe the process until the shock wave arises. Naturally, one wishes to extend the notion of the solution of equation (8.38). We shall look for the solution in the class of discontinuous functions, subjecting the character of the discontinuity to certain definite requirements corresponding to the physical nature of the phenomenon being described. Such a solution shall be called *generalized*.

We do not intend to construct the generalized (discontinuous) solution for the general case (8.40), and shall limit ourselves to the case of our physical example (1.33).

To the left of curve L (in domain I), the solution is constructed by the method described in subsection 1; this solution will be called classical. Further, taking L to be the projection of a certain new initial curve, we shall again construct a classical solution by the method of subsection 1. The question is to find what value of u must be taken along the curve L in order to construct the solution. To answer this question, we shall use physical considerations – the same conservation law of the quantity of matter which led us to the differential equation (1.33).

Denote by u_1 the limiting value of u on the line L (this line can be called the line of discontinuity); this value is obtained from the classical solution determined by the initial data, i.e. it is the limiting value from the inside of domain I (before the discontinuity). In order to construct the next smooth part, we must know the limiting value u_2 from inside domain II (after the discontinuity); this limiting value will be taken to be the initial one for the construction of the solution in the next step after the discontinuity.

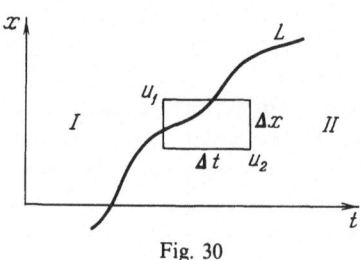

Fig. 30

To do this, let us write the balance equation in the following form (taking into consideration the fact that

$$u(x, t)=u_1, \quad u(x, t+\Delta t)=u_2, \quad u(x-\Delta x, t)=u_2\,;$$

see Fig. 30, where an infinitely small rectangle is shown; its left upper vertex is contained in domain I, infinitely close to line L, while the other vertices are in domain II):

$$-(u_2-u_1)\Delta x=[v(u_1)u_1-v(u_2)u_2]\Delta t.$$

Passing to the limit as $\Delta x \to 0$, $\Delta t \to 0$ and taking into consideration the fact that $\dfrac{\Delta x}{\Delta t}$ tends to the derivative $\dfrac{dx}{dt}$ of the function describing the curve (it is natural to call this derivative the velocity of motion of the discontinuity v_p), we obtain

$$v_p = \frac{v(u_1)u_1 - v(u_2)u_2}{u_1 - u_2}. \tag{8.49}$$

Knowing u_1 and v_p, we can use this formula to determine u_2.

Knowing u_2, we can again construct a smooth part of the solution up to a new curve of the same type as L. Such a curve, however, will not necessarily arise. We have seen above that the appearance of the first discontinuity was related to the fact that the characteristics "converged". No second discontinuity will arise if the characteristics "diverge" after the first discontinuity. Analytically, the presence or absence of the discontinuity can be established just as for the solution (8.48) above.

Bibliography

1. Bakhvalov, N. S.: Numerical methods, Vol. 1, Moscow: Nauka 1975 (In Russian) Moscow: Mir 1976 (In English)
2. Bogolyubov, N. N., Mitropolskij, Yu. A.: Asymptotic methods in the theory of non-linear oscillations. Moscow: Fizmatgiz 1963 (In Russian) Delhi: Hindustan 1961
3. Courant, R.: Differential and Integral Calculus, Vol. II. New York: Interscience 1955
4. Elsgolc, L. E.: Differential equations and calculus of variations. Moscow: Nauka 1965 (In Russian) Moscow: Mir 1970/73 (In English)
5. Erugin, N. P.: A handbook for the general differential equations course. Minsk: Nauka i tekhnika 1972 (In Russian)
6. Hartman, P.: Ordinary differential equation. New York: Wiley 1964
7. Kalitkin, N. N.: Numerical methods. Moscow: Nauka 1978 (In Russian)
8. Kamke, E.: Differentialgleichungen, 5. Aufl. Leipzig: Geest und Portig 1964
9. Kartashov, A. P., Rozhdestvenskij, B. L.: Ordinary differential equations and the foundations of the calculus of variations. Moscow: Nauka 1976 (In Russian)
10. Krylov, N. M., Bogolyubov, N. N.: Introduction to nonlinear mechanics. Kiev: Publications of the Acad. of Sciences of the Ukrainian SSR, 1937 (In Russian) Princeton, N. J.: Annals of Mathematical Studies no. 11 1943
11. Landau, L. D., Lifschitz, E. M.: Quantum mechanics. Moscow: Nauka 1974 (Course of theoretical physics vol. 3) Oxford: Pergamon 1977
12. Lomov, S. A.: Introduction to the general theory of singular perturbations. Moscow: Nauka 1981 (In Russian)
13. Marchuk, G. I.: Methods of computer oriented mathematics. Moscow: Nauka 1977 (In Russian)
14. Mishchenko, E. F., Rozov, N. Kh.: Differential equations with small parameter and relaxational oscillations. Moscow: Nauka 1975 (In Russian) N. Y., London: Plenum 1980
15. Nemytskij, V. V., Stepanov, V. V.: Qualitative theory of differential equations. Moscow-Leningrad: Gostekhizdat 1949 (In Russian) Princeton Math. Ser. 22. New Jersey: P.U.P. 1960
16. Petrovskij, I. G.: Lectures in the theory of ordinary differential equations. Moscow: Mir 1975 Englewood Cliffs, N. J.: Prentice-Hall 1966
17. Pontrjagin, L. S.: Ordinary differential equations. Moscow: Mir 1979 Moscow: Mir 1975 (In French)
18. Samarskij, A. A.: Theory of difference schemes. Moscow: Nauka 1977 (In Russian)
19. Smirnov, V. I.: A course in higher mathematics, Vol. II. London: Pergamon Press 1964
20. Smirnov, V. I.: A course in higher mathematics, Vol. III, Part. II. London: Pergamon Press 1967
21. Smirnov, V. I.: A course in higher mathematics, Vol. IV. London: Pergamon Press 1964
22. Stepanov, V. V.: A course in differential equations. Moscow: Gostekhizdat 1953 (In Russian) (Lehrbuch der Differentialgleichungen) Berlin: V.E.B. 1956, 1963, 1982
23. Vasil'eva, A. B., Butuzov, V. F.: Asymptotic expansions of the solutions of singularly perturbed equations. Moscow: Nauka 1973 (In Russian)

Subject Index

L. Hörmander

The Analysis of Linear Partial Differential Operators I

Distribution Therapy and Fourier Analysis

1983. 5 figures. IX, 391 pages. (Grundlehren der mathematischen Wissenschaften, Band 256)
ISBN 3-540-12104-8

Contents: Introduction. – Test Functions. – Definition and Basic Properties of Distributions. – Differentiation and Multiplication by Functions. – Convolution. – Distributions in Product Spaces. – Composition with Smooth Maps. – The Fourier Transformation. – Spectral Analysis of Singularities. – Hyperfunctions. – Bibliography. – Index. – Index of Notation.

Here is an unusually extensive presentation of the most significant techniques and results in the theory of linear partial differential operators. The field has undergone rapid development in the past 20 years, with the emergence of techniques involving pseudo-differential and Fourier integral operators – hence this updated and greatly expanded revision of Hörmander's previous treatment (Grundlehren Band 116).

By presenting the main methods, the author aims to make the vast literature available to a wider audience. The presentation leads the reader to areas of current research interest through a discussion of their classical roots, which is given in a modern spirit. Notes to the chapters indicate sources for the material and provide references to related work. Volume I is primarily directed to a broad range of students with an interest in analysis, assuming a basic knowledge of advanced calculus, integration theory, and some functional analysis. Going beyond previous books on distribution theory, Hörmander places a special emphasis on Fourier analysis, particularly its important microlocal aspects. These are also discussed in the wider framework of hyperfunction theory, providing a useful introduction (in the spirit of Schwartz distributions) to harmonic analysis and to the analytic theory of partial differential equations.

Volume II is a systematic study of partial differential operators with constant coefficients, and some of their perturbations. New chapters cover convolution operators, scattering theory, and methods from the theory of analytic functions of several complex variables. This volume (and the projected volume III) are more advanced, and directed to research workers and graduate students in mathematics and mathematical physics.

Springer-Verlag
Berlin
Heidelberg
New York
Tokyo

M. Braun

Differential Equations and Their Applications

An Introduction to Applied Mathematics

3rd edition. 1983. XIII, 546 pages. (Applied Mathematicial Sciences, Volume 15). ISBN 3-540-90806-4

Contents: First-order differential equations. – Second-order linear differential equations. – Systems of differential equations. – Qualitative theory of differential equations. – Separation of variables and Fourier series. – Appendix A: Some simple facts concerning functions of several variables. – Appendix B: Sequences and series. – Appendix C: Introduction to APL. – Answers to odd-numbered exercises. – Index.

This is the third edition of a well-known textbook.

From the reviews: ... this book is the most attractive motivated and well-written work of its kind known to me. Braun has integrated the teaching of elementary analytical methods of solving differential equations, numerical methods of solving differential equations, and the qualitative theory of differential equations with a variety of marvelously motivated, real-life applications that cannot fail to interest students...

American Scientist

This is an unusual, and a rather exciting book. It is unusual in that it is written in a racy style, completely unlike the rather dry factual style cultivated by most mathematical authors. It is exciting, both because it gives real examples of the power of differential equations applied to such diverse problems as the forged Vermeers and the incidence of epidemic diseases, and yet because it maintains quite a high degree of rigour. ... the book can only deserve the highest possible commendation. Buy it. Read it. Enyoy it – and you will be surprised how much you learn!

The Mathematical Gazette

Springer-Verlag
Berlin
Heidelberg
New York
Tokyo